- Radical Realism

• Also by Edward Pols:

The Acts of Our Being: A Reflection on Agency and Responsibility
Meditation on a Prisoner: Towards Understanding Action and Mind
The Recognition of Reason
Whitehead's Metaphysics: A Critical Examination of "Process and Reality"

Radical Realism

DIRECT KNOWING IN
SCIENCE AND PHILOSOPHY

Edward Pols

Cornell University Press

Ithaca and London

Copyright © 1992 by Cornell University

All rights reserved. Except for brief quotations in a review, this book, or parts thereof, must not be reproduced in any form without permission in writing from the publisher. For information, address Cornell University Press, 124 Roberts Place, Ithaca, New York 14850.

First published 1992 by Cornell University Press.

International Standard Book Number 0-8014-2710-X
Library of Congress Catalog Card Number 91-55531
Printed in the United States of America
Librarians: Library of Congress cataloging information appears on the last page of the book.

⊚ The paper in this book meets the minimum requirements of the American National Standard for Information Sciences—Permanence of Paper for Printed Library Materials, ANSI Z39.48-1984.

Contents

Preface ix

1 Introduction 1
   1. A Preliminary Look at the Scandal of Radical Realism: Direct and Indirect Knowing 1
   2. Contemporary Objections to the Distinction between Direct and Indirect Knowing 5
   3. The Oscillatory Relativism of the Linguistic Consensus 9
   4. A Negative and a Positive Purpose 16

2 The Scandal of Radical Realism 19
   1. The Reality Question: A Question in First Philosophy 19
   2. First Philosophy and the Questioning of Our Rational Birthright 26
   3. Radical Realism: The Cultivation of Rational Awareness 30
   4. Rational Awareness of the Relation between Reality and the Propositional 44
   5. Dismissing the Foundation Metaphor 50

3 The Linguistic Consensus   55
1. The Conflation of Language and Theory   55
2. Seven Dogmas of the Linguistic Consensus   57
3. Two Consensus Half-Truths about Science   72
4. Intimations of the Directness of Rational Awareness   77

4 Realism versus Antirealism: The Venue of the Linguistic Consensus   83
1. The Epistemic Triad and a Dominant Question about It   83
2. A Question Once Asked within the Consensus: What Is Experience Really?   87
3. The Consensus View of the Relations between the Members of the Epistemic Triad   89
4. Consensus Realists as Internal-to-the-Language Realists   91
5. Consensus Antirealists as Pseudo-Kantians   102
6. Why Antirealist Relativism Creates No Problem for Science but Many Problems for Other Fields   119

5 Radical Realism: The Venue of Direct Knowing   122
1. Two Functions of Rationality: The Function of Rational Awareness and the Formative Function   122
2. The Primary Mode of the Function of Rational Awareness   126
3. Three Scandals That Hinder Recognition of Primary Rational Awareness   132
4. The Secondary Mode of the Function of Rational Awareness   136
5. Rational-Experiential Satisfaction and the Natural Reflexivity of Rational Awareness   139
6. Dependence of the Formative Function on the Function of Rational Awareness   144
7. Indirect Knowledge and the Formative Function   146

6   Nine Theses about Science, Common Sense, and
    First Philosophy   152
       1. Introduction   152
       2. The Nine Theses   153

7   First Philosophy and the Reflexivity of Direct
    Knowing   175
       1. "The Science We Are Seeking": Bringing First Philosophy into Being   175
       2. Two Substantive Tasks of First Philosophy: Knowing Causality and Primary Beings Directly   181
       3. The Unity of the Two Tasks: The Causality of Primary Beings   190
       4. The Exemplification, in Rational Awareness Itself, of Its Own Findings in First Philosophy   207
       5. The Reflexiveness of Rational Awareness and the Movement from Primary Beings to Being   211

Index   217

- Preface

- A long time ago, when I was far too young to give it a name, let alone develop it adequately, I came to know something about knowing. It happened suddenly, on an occasion whose character in my memory is more precise than the date, which was, I think, some time in the fall of my senior year at college. The setting had nothing to do with the public—and frequently argumentative—side of philosophy, for I was quite alone; and it had nothing in it that seemed propitious for enlightenment. I was standing in the door near the fireplace of my study in Dunster House and looking into my bedroom, in which only the usual college furniture could be seen. A step or two into the room would have given me, from the high window, a view more likely to quicken a young mind: the tower of Lowell House, the tall campanile of St. Paul's, the spire of Memorial Church marking the Yard—at that time still an exuberance of elms—in the distance. But in my youthful epiphany my eye rested on only a worn bedstead and a chest of drawers.

What I enjoyed was a rational awareness (as I now call it) of ordinary things, but a rational awareness suddenly qualified and heightened by a surge in the intensity of the reflexive feature that is always native to it. And what that surge brought me was the assurance that we, the knowers, do not endow the known thing with the structure that comes through to us in our knowing it. In the language of this present book, I was rationally aware of certain things that were in no sense products of my act of rational attention; and I was also rationally aware (by virtue of the reflexivity inherent in that function) that those things were indeed thus independent of my rational attention. I emphasize the commonplace nature of the objects I was looking at because in the past I have often taken more interesting things—ourselves as persons—as examples of things attained by rational awareness. But any example will do: the point is not the example but rather the reality-attaining power of rational awareness. This youthful epiphany was nevertheless concerned with persons as well, for it was reflexive in nature and so had our own power of rational awareness as much at its focus as the dull furniture with which it began. As I said at the outset, I could not at that time have expressed what happened in the language I now use; yet it is not language that is authoritative but rather the reality which, asserting itself in our rational attention to it, insists that we should somehow find or make the language adequate to it.

What I attained in that moment (or what came to me) was in one sense commonplace, as commonplace as that furniture, but in another sense not. As time went on, I felt that I must be faithful to and develop what I had attained, youthful and incomplete as it no doubt was. It was many years before I came to grips with it in writing, but when I did, it played an important, though unacknowledged, part in my earlier books. Yet I think it has never been so well actualized as in the positive thesis of this book or so well related, in a critical way, to the current state of philosophy.

I am deeply grateful to three colleagues—David Braybrooke, Eugene T. Long, and Kenneth L. Schmitz—who undertook the considerable task of reading and commenting on the finished manuscript of the book. Their comments, which were thorough, illuminating, and encouraging, represent a generous gift of time and intellectual energy on the part of men who have much important work of their own to do. Their suggestions for improvements were not numerous, but all of them merited careful consideration. In responding to them I have introduced some changes that make for greater lucidity here and there.

An earlier version of Chapter 3 appeared as an article, "After the Linguistic Consensus: The Real Foundation Question," in the *Review of Metaphysics* 40 (1986): 17–40. Chapter 3, however, incorporates substantial changes, not the least of which is the disappearance of the reference to the foundation question. A later and related article, "Realism *vs* Antirealism: The Venue of the Linguistic Consensus," *Review of Metaphysics* 43 (1990): 717–49, appears here, with some changes, as Chapter 4. Both chapters belong to the negative, or critical, thesis of the book. The warm and understanding response to my work of my friend Jude P. Dougherty, editor of the *Review of Metaphysics*, has been of great importance to me over the years.

Roger Haydon, editor at Cornell University Press, has been a model of orderly intelligence, consideration, and tact since he first took an interest in the manuscript of this book. I count myself fortunate that he remained with me to help guide the book to publication. Through the various stages of production, the professional authority and kindness of Kay Scheuer, managing editor at the Press, have given me reassuring support. I am grateful to her and to her efficient staff.

EDWARD POLS

Brunswick, Maine

- Radical Realism

1

• Introduction

1 *A Preliminary Look at the Scandal of Radical Realism: Direct and Indirect Knowing*

• This book is about the nature and scope of rationality, more precisely, about its capacity to know reality: to know whatever in particulars and in the general nature of things is independent of our minds—whatever is in no sense dependent on any formative, or productive, powers rationality may also have. It is a topic so vital to the settlement of all other philosophical issues that we may well call it the first topic of philosophy. To take it up in a radically realistic spirit, however, is to create a philosophical scandal; so it is wise to give the reader some intimation of the nature of the scandal right at the beginning. Let us therefore consider at once a distinction that is central to radical realism: the distinction between two kinds of knowing, direct and indirect; for it is the claim that there is direct knowing—direct knowing of independent reality—that is a scandal to many philosophers.

Examples of direct knowing are easy enough to find; authenticating them in a rigorous way is another matter. You, as reader, now know directly the book you are reading, and you know directly any number of items in the place where you are reading it. Although any such commonsense object will do as an example just now, in the long run our knowing one another as persons will have an especial importance. Let us therefore introduce that example at once. If you and I were now talking together and not in one-way communication by way of the printed page, we should be knowing each other directly, and our consideration of the topic now in hand would be mediated by the direct knowing we should in that case enjoy or possess.

No doubt memory and language enter into and help make possible our knowing each other as persons. But those resources, brought to the place and moment of our conversation, do not cancel out the directness. Persons, as well as all the other commonsense things we might notice in the place of our meeting, are not remote and inaccessible things. Our knowing each other is not a matter of entertaining and assessing concepts or theories about persons—including theories that purport to show that there are no such things as persons. In other circumstances you and I might well decide to consider and debate theories of that kind. But our knowing each other in that imagined moment and place, though by no means exhaustive and by no means without obscurity here and there, is both rational and experiential. There, in that moment, we experience the place that environs us, and each experiences the other person; and the rational component of our knowing marches with that experience and is saturated with it, just as the experiential component is pervaded by our rationality. Many things about persons remain mysterious: there is, for instance, an inwardness to persons—a privacy—that is difficult of access; and there are many puzzles about the role physiology plays in our consciousness or in what purport to be moral choices. But that there are indeed persons and that we two are persons seems clear enough as we converse, unless philosophy has intervened to make us doubt what we seem to know so directly. In the past it was much the same, even in quite distant days when we first began to learn some of the things we bring to our present conversation. It was much the same too in those most distant days when we all first learned our language. Those persons who taught us our language, usually our parents, were experientially present to our rationality as they taught us.

There are three expressions that are roughly synonymous with 'direct knowing', at least as I shall be using them. They are 'rational awareness', 'rational-experiential engagement', and 'rational consciousness'.[1] Notice that all three of them are two-sided: each of them brings together the two realms traditionally known as reason and experience. Each of them insists on the interfusion or interpenetration of those realms, which so many philosophers have thought to be so distinct that a body of recondite theory must be devised to show how they can possibly be connected with and relevant to each other. I will argue that this is a mistake, and that the tight bond between the two realms is something so real and so authoritative that we need not prove it but merely call attention to it in an appropriate way. But an authentication of that kind is no easy matter, and we must set it aside for a while.

Indirect knowing is quite different. Many simple examples of it can be drawn from the world of common sense, most of them convertible to direct knowing in certain circumstances. But many examples of indirect knowing cannot be so converted. Thus, we cannot in principle know the interior of the atom in the way we know each other as we converse. Our knowledge of it is by way of a complex body of theory, mostly mathematical, by way of complex instrumentation, and also by way of the outcome in the commonsense world of experiments designed to test what theory predicts. If an example of this kind deserves to be called indirect *knowledge,* then there must be outside the body of theory entities or events that are at least somewhat like the entities or events we suppose or imagine to be there on the basis of the theory. Thus, if a supposed microentity, a quark, for instance, turned out in the

---

1. I use double quotation marks on (a) quoted material that is run in with the text, and (b) expressions that are unusual, strained, or in some sense dubious. I use single quotation marks on mentioned expressions.

long run to be merely an entity of theory—in effect, a useful fiction in the present stage of science—then it is not correct to say that we now have indirect knowledge of quarks. Again, according to today's cosmological theory, black holes exist in certain regions of the universe. That cosmological theory forms part of what is sometimes called grand unified theory. Right now there are several versions of that theory, but many people—perhaps more philosophers than working scientists—expect that some future version will for all practical purposes be a completed science. In order to assume that cosmological theory of this kind gives us indirect knowledge of black holes, we must also assume that black holes exist outside the theory. It is not enough that they exist as theoretic objects inside the theory, and that the theory is a good one in the sense that by virtue of it scientists are able to anticipate the outcome of various experimental situations in ways that would not be possible without the body of theory. There must also be certain regions of the universe where matter is even now falling into—vanishing into—that out of which not even light can escape. We can hardly be said to know something indirectly if that something does not exist but is a useful fiction.

But one thing is sure: if quarks and black holes are more than useful fictions, we cannot in principle know them as you, the reader, and I, the writer, know each other in our imagined conversation. What knowledge we have of them is and must remain something bound up with our knowledge of the theory. Perhaps our knowledge of the *theory* is direct enough— we shall look at that possibility in due course; perhaps there is an odd sense in which we—or at least a group of competent astrophysicists—may be said to know directly the structure of the theory and the concepts 'quark' and 'black hole' that are part of it. But knowledge of the quarks and black holes to which those concepts refer is, and must be, by way of the theory, and so it is indirect. There are, to be sure, other interpretations of the notion of reference than the one I am, for the moment, taking for granted; but those interpretations rule out indirect knowledge of real entities.

2 Contemporary Objections to the Distinction between Direct and Indirect Knowing

There are many academic philosophers today—perhaps a majority of the English-speaking ones—who would object that the distinction between direct and indirect knowing is not sound, for the simple reason that there is no direct knowing whatsoever, and so no direct knowing of persons—or, for that matter, of cats, dogs, squirrels, trees, flowers, and so on. All knowledge, they would say, is a function of the theories we hold and the languages we use. Those philosophers seem to think of the holding of a theory as a kind of language use. Curiously, they also tend to think of language use as a kind of theory holding. In short, they blend what, on the face of it, are separate notions or concepts, 'language' and 'theory'. To speak more bluntly, they confuse or conflate the two notions. Let us say that they make all knowledge a function of language-cum-theory, so much so that for them there is no way to get outside language-cum-theory and know *anything* directly.

But even that way of putting the matter does not quite get at what those philosophers are saying. It implies that persons, animals, and plants all have natures or structures which are quite independent of the knower; and that, by way of language and theory (thus, indirectly), the knower can nevertheless aspire to know these natures or structures as they independently are. That is not what they mean. Though these philosophers, or some of them, believe that there is *something* that is independent of language and theory, they do not believe that persons, animals, plants, and all the rest are part of that something: the only existence that all those things have is not independent of language and theory but is rather a function of language and theory. They concede that there is a coherent language—the commonsense one we use much of the time—which contains terms like 'person', 'cat', 'dog', 'squirrel', and 'tree'; and that there are occasions on which it is perfectly usual and sensible to speak of knowing persons and all the rest. But if we shift our language, say to that of physics, there will *be* no persons, animals, and plants for that language. In short, what I am supposing to be a direct knowing of persons and all those other commonsense entities is the result of the form-giving power of language and theory.

It is important to notice that philosophers who take this line are not merely questioning the independent reality of commonsense entities. Their claim applies also to the findings of science—to such things as the large organic molecules biophysicists talk about and the quarks and black holes physicists talk about. All supposed entities whatsoever have the same reality status, and that status is a function of language and theory. All entities whatsoever are *entities for a language;* each of them is said to belong to the ontology of a certain language and theory.

Such philosophers are usually called antirealists—less often, nonrealists or irrealists. The thought behind this name is that any so-called entity whose reality status is in fact a function of language and theory should properly be thought of as an antireality. Another way of making the same point is to say that all things that seem to be independent realities either to our senses or to our imaginations are merely appearances, or phenomena. And what produces such phenomena—produces them out of whatever bombards us with unknowable "stimuli"[2]—is the formative power we deploy in language and theory. As we shift from one language to another—say, from

2. Truly unknowable; these "stimuli" are not stimuli in the usual sense. The nervous system as a biological complex, individual neurons, and stimuli (in the usual sense) are also appearances (antirealities) according to this doctrine. These "stimuli" are postulated unknowables that lie behind all stimuli (in the usual sense).

the language of common sense to that of science, thence to that of art, that of morals, or that of religion—we get different kinds of appearances, or antirealities. There is no ground—at least no rational ground—for finding one kind more important than another. None provides a reality standard because all are antirealities, or appearances.

Philosophers of that kind belong to the linguistic consensus, which I shall examine in some detail in Chapters 3 and 4. That consensus, however, is not entirely monolithic about the reality question. An important subgroup insists that one body of language-cum-theory is privileged: that of perfected science, which they think we have good reason to aspire to, even though we do not now possess it. That body of language-cum-theory, they think, will indeed provide us in due course with what is real independently of language-cum-theory. The philosophers of this subgroup, then, are realists, but only with respect to that perfected body of theory. It is convenient to call them what many of them call themselves, scientific realists.

The general shape of the expected content of that perfected theory is already familiar to us, for who can escape the vision of the nature of things now placed before the educated public by so many skilled expositors of science in books, in the more sophisticated magazines, and on television? Central to it is the dominant cosmological theory, which postulates the origin of the universe in an event whose common name—the "big bang"—is less dignified than so distinguished a thing deserves. It is thus assumed that this event will belong to the ontology of whatever grand unified theory finally emerges as the real comprehensive unified theory and no mere aspirant one. It would be consistent with the spirit of scientific realism of that kind to say that such an event is an extratheoretic one, known indirectly by way of a perfected body of language-cum-theory. But to say so is to claim that perfected science would have given us, at long last, something that was a true ontology and not just the "ontology" of a language. In any event, it is supposed that the comprehensive unified theory which in due course succeeds in laying all this out—from the beginning until now, and from now on to whatever infinite expansion, whatever ultimate extinction or regeneration may be—would leave no secrets or mysteries. It would tell us the way of things and moreover assure us at the same time that it *was* the way of things rather than some useful creation of the scientific imagination that enabled us to predict the course of experience but was not to be taken literally—like the crystalline spheres of Aristotle or the epicycles of Ptolemy.

But how shall we know that we have finally moved from mere "ontology" to ontology? How shall we know that we at last possess a body of language-cum-theory that gives us (indirect) knowledge of the extratheoretically real? Not, surely, by even an extraordinary increase in empirical prediction and control, for even the most explosive expansion of that kind can easily be given an antirealist interpretation. The real criterion for the reality-attaining power of the language-cum-theory of completed science is that nothing should resist its explanatory scope. No apparent reality—nothing that purports to be known directly, like persons, animals, and plants—should remain obdurately unsubdued by that completed web. Most momentous of all, the knowing subject must not escape that web: the intelligence, consciousness, morality, and purposiveness of the human beings who devise the body of theory on the one hand and the philosophical defense of it on the other must not be left unreduced. If that (apparent) ontic level should remain intractable—if only some other mode of explanation could deal with it—then that body of language-cum-theory would not in fact be the ideal one by virtue of which we know the real through and through.

As we shall see, the disagreements within the linguistic consensus are less important than their agreement about the role of language-cum-theory in human knowledge. And this agreement means that there is complete accord within the consensus about the impossibility of direct knowing: for the consensus, there can be no direct rational awareness of independent reality. Against that consensus, I ask you to take seriously—pending later authentication—the possibility that there is indeed direct knowing and that the distinction I have made between direct and indirect knowing is accordingly sound. It would follow that there are entities having natures that owe nothing to any supposed formative power of language-cum-theory, that we can know some of them directly, and that we can know others only indirectly and by way of theory. On the other hand, some things that we now think we can know indirectly may not really be there.

3 *The Oscillatory Relativism of the Linguistic Consensus*

It is usually supposed that there is at least one important disagreement within the linguistic consensus: antirealists tend toward relativism; scientific realists do not. But I think we shall find that at a deeper level both positions exist in an uneasy oscillation between relativism and absolutism, and that this oscillation makes for an overarching relativism that may be the most significant contribution of the linguistic consensus to late twentieth-century culture.

Let us consider antirealism first. It is relativistic in a sense that seems to be beyond dispute: it maintains that the thing known is relative to the shaping power that makes it, qua known. Even when we deploy our rationality in the precise activity of science, antirealists say, we can do no more than produce concepts, categories, hypotheses, models, laws, equations, and the like and then use them to order and anticipate our experience. For antirealism, strictly interpreted, even the latest comprehensive unified body of scientific theory has no more profound reality function than that. Like all bodies of theory, that comprehensive one would have been *formed by* the linguistic-cum-theoretic power of human rationality, and so, whatever its predictive and organizing virtues might be, it would remain distinct from and in some important respects different from the reality of which it purports to give us knowledge. As for experience, to which that body of theory is relevant, that too is a product—or at least a partial product—of the formative power of rationality, and so it cannot put us into any more intimate touch with reality than the body of theory can.

A relativism of this kind is compatible with a rigorous standard for truth in science. We might, for instance, suppose ourselves to be dealing, in science, with something merely phenomenal but might nevertheless insist that we can apply rigorous truth standards within that phenomenal realm. Just so, Kant had a precise standard for scientific truth, even though he claimed that science could deal only with the phenomenal. But, as we saw earlier, antirealists do not suppose that our rationality forms antirealities only in the realm of science. Their thesis ranges over all our rational activities. There is thus a deeper and more troubling relativism lurking within their doctrine. The source of this relativism is the diversity of interests that move rationality and the consequent diversity of its formative response to the ineffable reality with which it is somehow in touch. Thus, whether rationality deploys itself in social science, in practical moral choice, in philosophical theory about moral choice, in creative art, in theory about creative art, or in one of the manifold other activities we group under the rubric of common sense, it must produce an antireality or antirealities correlative to that mode of deployment.

Many of these activities in which rationality may be deployed were once understood in a realistic way—that is, they were thought to involve a grasp on the part of rationality of something whose reality status was not dependent on rationality. Accordingly, there was a presumption that truth was in principle attainable—at least some of the time—in such activities. But in the present antirealistic setting, what is called truth in one of these deployments may be falsehood from the point of view of some other deployment; so the notion of truth must be dismissed in favor of a more flexible standard. Both critics of antirealism and antirealists themselves have felt that this linguistic relativism, as it is sometimes called, poses a grave problem for the doctrine. At the very least, it raises questions about how we are to compare the importance of what is said when one of our interests rules (say that of science) with what is said when another interest rules (say that of morals).

If we consider only morals, the antirealist claim must apply not only to our particular choices but also to the correlative general moral doctrines that bear some relation to particular choices. Thus, the performance of a particular act because, in some situation of apparent moral choice, the act seems to be good, appropriate, right, or obligatory, must in fact be ascribed to the formative power of rational evaluation deployed in that situation. What appears to be really good or really obligatory in the act is in fact the consequence of the formative rational power of the person who responds to the "goodness" or the "obligation." But since no person acts in utter detachment from a community, no moral decision is made in complete isolation from some correlative moral doctrine or attitude that is shared with others. What is shared may well be a system of ethics that purports to be based on something in the real nature of things that makes it right or good for us to act in a certain way. Antirealism, however, tells us that the persuasiveness of any such doctrine is also an outcome of the deployment of the formative power inherent in our rationality. This embedding of particular choice in a general attitude or code is usually ascribed by antirealists to what they call a linguistic community or community of discourse. The formative power of rationality is thus taken to be a *linguistic* formative power. Goodness and soundness are thus not determined by something *real* in the nature of things; it is rather ourselves, as a linguistic community, that make a certain action good or a certain ethical system sound. The apparent "real" correlative of the linguistic-cum-theoretic act of ascription is in fact engendered by the ascription itself.

Art is a special case, for there is an obvious and undisputed sense in which the arts produce something that is not real. There has been considerable antirealist writing about art in recent years, some of it, I think, directed toward making this obvious feature of art a precedent for all rational activity. Long before this contemporary controversy, however, many writers and artists and some philosophers have made the claim that the nonrealities, or antirealities, of art may be vehicles for penetrating ordinary reality and reaching a deeper reality—the real being of Plato. Romantic art theory is full of such claims, and so is twentieth-century theory of abstract art. It is, however, clear enough that a relativistic antirealism must reject claims of that kind.

Besides the avowed antirealists we have been considering, there are some whom the rest of the world would want to call antirealists but who themselves repudiate that label. If they accepted it, they think, they would be acquiescing in the complex claim that there is indeed something both independent of rationality and forever inaccessible to it; that, despite its inaccessibility, it is an appropriate and ultimate cognitive criterion for us; and that, even though we can never confront it and see it for what it is, it and it alone deserves to be called real. In their judgment, this doctrine is not even intelligible, and so they urge a shift in the meaning of the term 'reality' in a relativistic direction. Let us, they say in effect, include henceforth in the notion 'real' a concession about the mind dependence of anything we may be said to know. Henceforth let all the internally coherent things that are produced by the formative power of rationality be called real. Such things, whether merely linguistic-cum-theoretic or else experiential in a sense that depends upon our language-cum-theory, are the only ones we can possibly concern ourselves with. And indeed, so they claim, our whole commonsense world is mind-dependent in precisely that sense. Why, then, should not all such things be entitled to the honorific overtones of the word 'real'?

Let us call this altered sense of the term 'real' a Pickwickian sense, a shift in meaning that we can also arrange with the help of transformatory double quotation marks—"real." So the group usually called antirealists (or nonrealists) will now be understood to include some Pickwickian realists, or "realists." The relativism we have already glanced at in considering avowed antirealists is also present in Pickwickian realism. But while avowed antirealists sometimes find relativism an embarrassment, Pickwickian realists tend to glory in it, taking it as a sign of the exuberant creativity of our rationality: the more "realities" the better, whether art, ethics, the law, or even theology is our theme.

The maneuver of shifting the normal sense of the term 'real' in a relativistic direction is at least as old as the Greek sophists. The story is a familiar one, but it is worth a glance anyway—if only because it reminds us that another important word lies behind our Latin-based 'real'. The word 'real' goes back to the Latin word for thing, but it comes into general philosophical use only in western European philosophy. When the Greeks dealt with what today we might call the reality question, they usually employed some inflectional form of the verb 'to be'. Thus the Platonic expression *ontōs on*, credibly translatable into English as 'really real', consists in the Greek of an adverb and a participle both of which are forms of the verb 'to be'. That verb, then, carried the implication that what *is* is in no sense dependent on the knower who knows it, and perhaps the further implication that the whole point of knowing is to attain precisely that which *is* thus independent of knowing. A philosopher of those times who might today be called a realist would have taken it for granted that what was ontologically independent of the knower was both accessible to and cognitively attainable by the knower. Such a philosopher might well have claimed, "I *know* what *is*." So the relativistic maneuver in the days of the sophists was sometimes executed by conflating the normal use of such a word as 'seem' with the verb 'to be'. Thus Plato represents the sophist Protagoras as claiming that what *seems* to a man *is* to him (*Theaetetus* 177C), that each of us is the measure of the things that *are* and the things that are not (*Theaetetus* 166D).

Whether consensus antirealists are abashed or complacent in their relativism, they do in effect make one absolutist—that is, realistic—claim. It is a momentous one, for it is a claim about the nature and condition of human rationality.

Here, they say in effect, here in the very propounding of the doctrine of antirealism, we reach truth; here we get at the real, here we attain things as they are. I say "in effect" because, although there are plenty of explicit claims made by antirealists about the conditions under which human knowledge operates, those antirealists who are also avowed relativists show themselves to be well aware that they must avoid making explicit truth claims or reality claims. If we challenge them about this paradox, they concede that the position of relativism can be asserted only in a relativistic way: it is nothing more than the mode of discourse *about knowledge* that is found comfortable by the relativistic community of discourse. On the other hand, anyone who disagrees with this consensus position is assigned, by virtue of this imagined reply, to another language community; and this assignment in effect reinstates an absolutist claim about the supposed predicament of rationality.

Whether the claim on the part of antirealists that the human cognitive condition *really is* relativistic is expressed explicitly or conveyed indirectly by some maneuver designed to turn away the charge of paradox, it stands out as a paradoxical dispensation from that condition. Even if it should be the only exception, it is nothing if not absolutist and realist: human rationality, it says in effect, is doomed to deal with antirealities—mere rationally formed "realities"; and it is so doomed just because its epistemic condition *really is* thus and so.

In our first glance at the realist wing of the consensus we seemed to be far from the relativism of the antirealist wing. That, however, is an illusion, for the absolutism of scientific realism exists in an uneasy oscillation with its own kind of relativism. To take the most obvious point first, scientific realists take it for granted that any body of language-cum-theory other than that of science functions to produce antirealities. They are thus relativists with respect to all disciplines and attitudes other than those of science. Practically speaking, then, relativism encroaches on the lives of scientific realists, for they are not working at science all the time and so often find themselves engaged in activities that they cannot in principle take to be concerned with reality in the same serious sense in which they understand science to be so concerned. And some of these activities—for instance, what purports to be moral choice—seem to be of immense importance, not only to the rest of humanity but to scientific realists as well, for that category no doubt includes as many persons of high moral integrity as any other human category.

But this relativism also encroaches on their own philosophical doctrine, for their accounts of the way knowing functions differ in no significant way from the account I have already attributed to the antirealist wing of the consensus. For the consensus in general, knowing is the providing of propositional structures that meet certain standards; and, for the realist wing no less than for the antirealist one, those structures are thought to emerge from the formative power of language-cum-theory we deploy in response to theoretically postulated "stimuli" whose very nature is ineffable except by way of another equally formative response.

How, then, can the scientific spirit transcend the fate of rationality in those lesser and admittedly relative activities, those activities to which the whole question of reality is irrelevant? How can *scientific* rationality, while deploying itself in precisely the same way that ensures the merely relativistic nature of these nonscientific pursuits, rise above the relative to the attainment of reality? That the outcome of this doomed-to-relativity activity will in the long run be the attainment of the absolutely real is the nub of their paradoxical claim. We glanced earlier at their criterion for success: the elimination from the ultimate body of theory of any residue of the human and hence any residue of humanity's relativistic predicament. But at the same time they tell us that as we try to accomplish this task we have no rational capacities at our disposal that are not defined by this very predicament.

It seems clear enough, then, that the polar doctrines of the linguistic consensus are inherently unstable. The relativism of the antirealists goes over into absolutism, from which it is then thrown back into relativism. The absolutism of the scientific realists, on the other hand, acknowledges the supposed relativistic epistemic predicament of the nonscientific attitudes and enterprises of humanity but has no alternative epistemological doctrine to explain the activity of science as it is now conducted—or, for that matter, the activity of philosophy as it is now conducted. From this paradoxical situation the scientific realist recoils into an arbitrary absolutism with respect to a perfected science from which the knowing subject shall have expunged itself qua knowing subject—shall have done so, moreover, even though, as inquiring subject, it must operate from the same relativistic epistemic predicament.

This oscillatory relativism is not the only reason for the instability in rationality's self-image in the latter part of the twentieth century, but it is nonetheless an important one. That self-image is nothing more than the image of rationality entertained by people like the reader and myself, and by a great many other thinking people who have no deep interest in academic philosophy and little or no technical knowledge of a realism-antirealism debate within it. This self-image is unstable in the sense that educated people in general tend to oscillate between antirealism and realism—hence, between relativism and what we are here calling absolutism, having recourse to one attitude in a certain situation or mood and to the other when that situation or mood changes. It is important to notice, however, that an oscillation of this kind is itself relativistic in spirit. Despite the influence of science, the chief characteristic of twentieth-century rationality is relativism.

4 *A Negative and a Positive Purpose*

From this preliminary confrontation between radical realism and the linguistic consensus, it should be clear that in this book I have a negative as well as positive purpose. My negative purpose, which dominates Chapters 3 and 4, is to expound and discredit the doctrines about language and its role in the exercise of our rationality that define the linguistic consensus, and to discredit with them the prevailing relativism to which the linguistic consensus has made so potent a contribution. But my negative purpose is completely subordinate to the positive one. The linguistic consensus has imposed an adamantine orthodoxy on all discussion of the nature and scope of our rationality, and while its hold remains unshaken the intellectual community will continue to fail to notice that our rationality has a radically realistic birthright.

My positive purpose is to draw attention to the experiential engagement of our rationality with reality and, in doing so, to show that the function of language in the life of rationality is not what the consensus claims it to be. To anticipate: although language is essential to our construction of theories and doctrines, it does not function constructively, or constitutively, in other cognitive transactions and so does not make a direct rational-experiential engagement with reality impossible.

I say "draw attention to" deliberately. The phrase is meant to suggest that philosophy has another task besides the creation and analysis of doctrines. If I can make it clear, in the course of this book, just what that task is, the reader may be less likely to suppose that the book is an expression of philosophical conservatism. Quite the reverse: I am not proposing that we should turn away from the linguistic consensus and return to some metaphysical-cum-epistemological position of the past from which the consensus has somehow strayed. The realism that will in due course be put before the reader is radical and, perhaps, unprecedented—unprecedented, to be sure, only as a philosophical position, for if radical realism is sound, what it seeks to draw attention to is a familiar, if unacknowledged, factor in all our rational occasions. If I am right about this last point, my positive purpose cannot be accomplished by a mere contrast with the linguistic consensus. The linguistic consensus is merely one of today's instances of a long-standing failure of philosophy to acknowledge and accept the scandal of our capacity to know the real directly.

2

- ## The Scandal of
  Radical Realism

1   *The Reality Question: A Question in First Philosophy*

- The question of the relation between the human mind and that in particulars and in the general nature of things which the human mind has in no sense made has an importance that transcends the issue of our confidence in physical science, momentous as that issue no doubt is. The question is central also to our estimation of the significance and worth of morals, religion, and art, and so it is central as well to the fate of any society bound together by more than science. From this wider perspective, the realism-antirealism debate in today's philosophy of science is a provincial matter, and the senses of the term 'realism' defended or attacked in its skirmishes are inadequate to our true condition as knowers and so also to our true condition as human beings, with all that involves in the way of feeling, valuation, purpose, and action.

  How do we set about answering a question of that kind? I say "we," because in the nature of the case the investigation is not of that artificially private kind in which the very existence of our fellow human beings is alleged to be part of the problem. And we know this, not because we are using a received language that somehow makes it so, but rather because we are rationally aware of living with our fellow human beings, who have invented language and now continue to sustain and enrich it, and because we are thus rationally aware of language as but one feature of that common life.

Notice that these remarks are a function of the ontic level of the person—that they both call attention to certain features of that level and rely upon the same level to do so. There is nothing offensively circular here, for the ontic level in question is naturally reflexive; in any event, there is no way we can avoid relying on it, no matter what recondite business we have in hand. It comprises not just the making of claims like those I am now making, and not just all the unbuttoned doings of an easy-going common sense; it comprises as well every kind of precise rational activity, including whatever goes into the production of the most refined and searching criticism, the most complex and ingenious theories, and the most austere and metaphysically noncommittal formalisms. Even an attempt to show that there is no reality to the ontic level we are now inhabiting together, that it is just an apparent level—something perhaps linguistically formed out of we know not what—even that attempt must rely on the same level to make its case, must avail itself of all the nuances of that level, all its tricks and short cuts, all its assumptions.

How then do *we* set about answering the original question? It is a question about the real (that in particulars and in the general nature of things which the human mind has in no sense made); but it is also about the human mind and its capacities and limitations. That makes it a strange question indeed, for in the very posing of it we must take the mind to belong to the real about which we raised the original question. We want to get at a function of our mind that it cannot have made, whatever else it may have made—precisely the *real* function that does or does not permit us to get at the *real*. Whether we are realists or antirealists, nonrelativists or relativists, the terms of the debate make us suppose that we are able to do this.

It is thus paradoxical to propound an antirealism (nonrealism) or a relativism without putting a limitation on its range or qualifying its authority in some other way. For to claim to be in a position to say that one of these doctrines is unqualifiedly true about our epistemic condition, and moreover true in a sense that transcends any formative powers of our own rationality that might operate to make it seem true, is to take for granted, but not to acknowledge, a reality-attaining power on the knower's part that is authentic enough to authorize such a claim. The philosopher who asserts the unqualifiedly relativistic doctrine thus tacitly qualifies it: that philosopher, at least, is an exceptional case. If, on the other hand, antirealism and relativism were limited to certain epistemological circumstances, and if it were conceded that we can assess what the circumstances are by relying on radically realistic powers that operate under their own limitations, that prima facie paradox would disappear. In that case we should be in a position to supplement the limitations of our finite realistic condition by constructs of many kinds, and some of these constructs might then be given an antirealistic interpretation without affecting the underlying realism in the least. But the antirealisms that are commonly propounded are so unqualified and unlimited that the paradox persists, and this tempts us to turn aside and concoct a knock-down argument that would convict their defenders of logical confusion. Let us, however, resist that temptation, for whatever merit such an argument might have, it would surely not persuade someone who is already convinced of the correctness of unqualified antirealism.

We shall do better, at this stage of the game, to attend instead to a parallel between the reflexive intricacy I have attributed to the realism question and the reflexive intricacy that has always characterized an ancient and as yet unsatisfied philosophical appetite for the real: that which expresses itself in metaphysics.[1] Philosophers who manifest that

1. I mean to exclude all those antimetaphysical philosophies, and they

appetite in an unambiguous way already assume that our rational birthright includes an engagement with the real; this engagement, they say in effect, is never quite lost, no matter what doubt, obscurity, and confusion we sometimes find ourselves in. In the minds of these philosophers, the aspiration to the real is already to some degree fulfilled, and much of their philosophy is accordingly directed to what still remains unclear and incomplete in our intercourse with the real. As they see it, the incompleteness can be remedied and the obscurity dispelled only if our insight into our rational engagement with the real is itself deepened and perfected. As they work at this task, they exemplify, although they do not explicitly acknowledge, the reflexive intricacy of the realism question. Philosophers of that tradition thus practically acknowledge by their pursuit of the most important of epistemological questions that our realistic birthright is not merely at our disposal to be passively enjoyed and exploited but must be constantly justified, renewed, and deepened by virtue of our own reflexive attention to and cultivation of it. Only if we manage to become rationally aware of that *feature of reality* which is our own rational power to attend to, recognize, and absorb the real of whatever kind can we deploy that power in the way it needs to be deployed. There is no legitimate approach to reality unless

---

are many, whose originators insist on borrowing the term 'metaphysics' to characterize their own enterprises. Hume, for instance, tells us that his own antimetaphysical work constitutes true metaphysics. And Kant in effect does the same thing. Consider, for instance, the full title of the *Prolegomena*, which posits a true science of metaphysics as the eventual goal of critical philosophy: *Prolegomena to Any Future Metaphysics That Will Be Able to Present Itself as Science*. Those interested in the oddities of philosophical terminology will find at least two senses of the term 'metaphysics' in the *Prolegomena*—one for a study Kant rejects and one for the perfected critical philosophy he aims at—and there may well be a third sense. It is also clear enough that the latter part of the first critique has more in common with the rejected study than Kant intended. This curious affection for the term 'metaphysics' persists into our own day; at least some influential analytic-linguistic philosophers have called their own work metaphysics. See, for instance, Peter Strawson, *Individuals: An Essay in Descriptive Metaphysics* (London: Methuen, 1959); John Wisdom, "Philosophy, Metaphysics, and Psycho-Analysis," in his *Philosophy and Psycho-Analysis* (Oxford: Oxford University Press, 1953).

by virtue of it we can show just how our experientially engaged rationality does indeed engage the real in that or in any other sense.

It is not surprising that we cannot point to any philosophy that is acknowledged to have achieved this double—and moreover tormentingly reflexive—task. Nor is it surprising that when we examine any metaphysical writing—whether something written before Hume's criticism and Kant's consequent retrenchment, or something more recent that purports to answer, circumvent, or co-opt those great critics—we are struck by the deficiencies of what we may now call the epistemological correlative that the philosopher offers in support of the metaphysical doctrine. More often than not we are inclined to feel that, if our rational powers can in fact engage the real—or engage whatever it does engage that might fall short of the real—only in the way laid down by the epistemological correlative, we have little reason to believe that the real does indeed possess this or that feature attributed to it by the corresponding metaphysics. A classic case of an epistemological correlative that undercuts a philosopher's metaphysics is Descartes's doctrine of the representative role of the ideas possessed by the thinking being: a mind that could know only ideas that purported to represent the real would be in a poor position to know what is the case about its own cognitive capacities and so would be in no position to excogitate a respectable epistemology. We shall in due course be in a position to see that one reason for these difficulties is that, speaking in terms of the division of philosophy into the traditional fields, the metaphysical part of a philosopher's doctrine usually has the status of a *philosophical theory,* and so also with the epistemological part.

We may sum up this line of thought in this way. The metaphysical aspiration that goes back to the beginnings of philosophy and persists, despite all criticism, into our own day has been an aspiration to, and also a need for, the real in a most comprehensive sense—not just a sense adequate for natural science. Those who have felt the aspiration assume that it cannot be satisfied unless we, as knowers, can also know and express something about that reality which is our rational capacity to attain the real cognitively. Their confidence in this capacity is so unfailing that they have continued to rely on it—continued to act as though they did in fact have that capacity—even when their own epistemological accounts of it (which in general have the status of theories about it) made it seem that it could not in fact do what they wanted it to do.

I suggest that we can know the real in the sense intended, but not so far justified, by the metaphysical tradition; and that our first philosophical task is to so intensify and perfect the reflexive component of our rational awareness as to produce such a justification of its own authenticity. To put the matter in rather flat-footed methodological terms, the production of a sound metaphysics requires an adequate epistemological correlative. That in turn requires an *attention to* our own rational capacity rather than the production of one more *theory about it* that hides what was right there and accessible to our reflexive vision and substitutes for it some ingenious mechanism or some set of intermediaries which, once the internal logic of the substitute is developed, turns out to exclude the very possibility of that capacity's doing what—all the while—we are relying on it to do.

The thought that metaphysical and epistemological *theories* are part of the trouble suggests that it might be wise to disembarrass ourselves at once of the traditional name for an enterprise of this kind—'metaphysics'—for the name itself distracts us from the simplicity and concreteness we must preserve and nurture at the heart of our subtle task. It is a historical accident that gave us the expression *ta meta ta physika* for certain writings of Aristotle—'the [treatises] after the [treatises about] physics'—and though in its literal meaning it is innocuous enough, merely recording the arrangement of Aristotle's text by later editors, the literal meaning has long since been lost sight of. The *meta* is now taken to mean 'beyond', and with this change in meaning there goes a widespread notion that the discipline is a highly theoretic or speculative one, one whose connection with the human condition consists only in the obligation to interpret our experience in the light of theories whose primary mission is to tell us about physical things that are hidden from our experience and also about whatever may lie quite beyond the physical.

Fair enough, we might concede, that the name should encourage that notion, for many metaphysical doctrines do indeed fit that description; after all, the extension of the sense of 'metaphysics' from 'treatises after the treatises about the physical' to 'treatises that concern what lies beyond the physical' must have come about in part because readers were attentive to what was actually being said in so many books of metaphysics. My point, however, is that we need a change in the orientation of the discipline, and that a change in the name may help us bring that about. A metaphysics adequate to deal with the reality question will certainly be a subtle enough affair to please even the most austere taste, but it may well turn out to be a much more down-to-earth enterprise than the theory-ridden discipline of the past. If we manage to turn metaphysics in that direction, the discipline may eventually acquire a nonspeculative authority that will permit it to assess the nature and limitations of speculation and theory—whether exercised in the service of speculative metaphysics, science, or common sense.

There is a better name at hand for an *instauratio magna* of this kind—the one Aristotle himself used: 'first philosophy', *protē philosophia*. It suggests all the right things: the philosophy upon which other kinds depend; the philosophy that must be done either before the rest of our philosophy or else simultaneously with the rest, so that the renewing impulse is always there in each phase of the development, lending its authority and assurance to each of the rational acts that produce the philosophy. First philosophy in this sense remains, however, only a discipline aspired to, not one that we can point to as fully actual. Aristotle has other names for it as well; he also has a striking expression that is not a name at all but merely a most felicitous reminder that the discipline is still *in potentia*—'the science [or knowledge] we are seeking' (*Metaphysics* 983a21–23; 995a24).

2 First Philosophy and the Questioning of Our Rational Birthright

Philosophical reflection about our capacity for satisfying the appetite for the real has always been colored by the thought that we may nevertheless have to make do with something short of it. This is an innocuous enough thought—no threat to our confidence in our rational birthright—if it merely contrasts our aspiration to reality in its deepest and most comprehensive sense with our actual achievements; the latter, whether in physical science or in any other sphere, are at best progressive and never complete. The thought becomes paralyzing only if it represents our best rational achievements as not engaging the real in any sense.

Let us bring this thought closer to today's controversies by putting it into the form of a question, a question that raises the possibility that there is something quite spurious about our rational birthright and hence about our cognitive aspiration to the real. Can we ever know anything that is real in the sense of being quite independent of our rationality—independent, that is, of any formative, creative, or shaping power our rationality may possess? Philosophers who persist in thinking that the achievement of a sound first philosophy is possible remain obdurately convinced that what purports to be our rational birthright is authentic, despite all the obvious limitations on it. They have therefore faced up time and again to the apparent threat of that question—have done so, in fact, whenever they have coped either with doubt or with the problem of distinguishing between opinion and knowledge. In doing so they have in effect, although seldom explicitly, judged the question itself to be spurious—spurious in the sense that the possibility of an unqualifiedly negative answer can be dismissed immediately. Rightly understood, their responses seem to say, the question is merely an incitement to fight our way through the doubt, error, and uncertainty that admittedly belong to the finitude of the human condition. Radical realism makes common cause with that judgment. It deals with the reality question by calling into play a reflexive exercise of the same rational-experiential engagement with the real that enables us to entertain the question in the first place, to perceive that it has the linguistic form of a question, and to distinguish it from the situation (and there will always be *some* situation) within which the question is entertained. Facing the reality question so, we disarm it at once: we continue to take it seriously as a question about how we can pronounce upon our engagement with the real; but as a question about whether we are so engaged, we summarily dismiss it.

The question—call it the question about our capacity for an experiential and rational engagement with the real—gives rise to others. The most important one, important because it must be taken seriously even by someone who is utterly convinced of the authenticity of our rational birthright, concerns what I called the comprehensiveness and generality of the real we aspire to. It is this: supposing that we can know the independently real, are there some modes or levels of reality that are more important than others? If so, how much of reality can we know? How far does our cognitive reach extend? This raises another question, one that seems important to me but might be rejected by many contemporary writers because it supposes a sense of 'know' they have dismissed in advance. Are some few independently real things known directly, so that we can in principle use them as fulcrums for knowing indirectly many things that are now, and perhaps forever, hidden from us or otherwise inaccessible to us? The most common example of such hidden things are the microentities whose status figures so prominently in the realism-antirealism debate within philosophy of science. Another and related question is even more remote from today's controversies in academic philosophy: are all real things joined in a unity, a unity so universal that the unity of each particular real thing is an instance of it? And if that should be so, is our access to that unifying principle direct or indirect? This last question concerns what Plato called real being, or, in another translation, the really real; for many writers in the past, and for some few even today, this question ramifies in others about the status and cognitive accessibility of God and the Good.

The Latin-based term 'reality' has been well established in philosophy for a long time, and it would not be prudent to recommend that we give it up. Still, the term 'being' has precedence: the expression 'really real', as noted earlier, is made up of two forms of the verb 'to be' in the Greek. The cognitive appetite for the real that gives rise to these last few questions may thus be reexpressed in a form that goes back to the ancients: it is a need for Being in general, or for Being as Being. This is not merely a need for a very general *concept* of Being or for the *form* 'Being'. Nor is it merely a need to know what "ontology" goes with the use of a certain language. Call it, rather, a rational appetite for Being as one and universal, and as present in each of the particular things we are rationally aware of. If this appetite is authentic in the sense that there is something, rationally and experientially accessible to us, that can satisfy it, then our interest in it by no means diverts us from whatever particular real things—or whatever particular aspect of real things—we might be concerned with, for the particularity of each particular thing would in that case be intimately involved with the unity and universality.

It is the thesis of radical realism that this appetite is authentic; that the first steps we take toward satisfying it assure us that our knowing is never concerned with utter particularity; and that no experience of a particular, no particular act of rational attention, and no knowing of a particular takes place without a counterpart hold on Being in general. To make that point and the more difficult one that in the case of Being its features of unity and universality are in mutual support, let us take them together and speak of them as the

U-factor—a factor present in every particular being and indeed essential to the very particularity of it. If this claim should be true, then whatever unity we find in any particular being is not supplied by our own cognitive faculties; and whatever systematic unity we find in our knowing in general does not go back to some synthesizing power we possess as knowers. In our knowing, we do not deal with a mere multiplicity of discrete unities, nor is the unity and universality we find in knowing a function of a linguistic-cum-theoretic network imposed by our rationality.

To push the matter to the level of first philosophy as I am now doing is to express dissatisfaction with a variety of contemporary doctrines, among them the antirealist one that when we are pursuing that epistemic ideal we are doing so by imposing a linguistic-cum-theoretic unity upon ineffable stimuli that certainly lack *that* unity, thus making them into a world with an inner coherence that makes prediction and control possible. But the impulse toward first philosophy is not generated only by a dissatisfaction with philosophy of science; the impulse was there long before there was much science to philosophize about. It is generated also by the conviction that our own best completion as persons requires that the whole pattern of our habits, feelings, and valuations be consonant with the U-factor of Being. If radical realism is sound, that consonance becomes a more attainable ideal than has been supposed.

Radical realism calls our reflexive attention to something philosophy has not so far managed to attend to with enough steadiness and clarity: the fact of our rational-experiential engagement with the real, and its potentiality to serve as a source of insights to replace the unsatisfactory philosophical bodies of theory that have hitherto supplied us with our only notions of such matters as identity, body, mind, causality, and substance.[2] Radical realism insists, moreover, that the conviction that some such happy outcome is at least possible has

2. For the notion 'body of theory', see Chapters 2 and 3.

been subliminally present in all past philosophy that has tended toward first philosophy. Philosophers working in the tradition of first philosophy have persisted in supposing that we are in fact cognitively engaged with reality, even though they have not in fact been able to show that that is so, and even though they have not made it clear enough that there will be no progress unless that defect in their philosophy is overcome. Radical realism makes common cause with that tradition but nevertheless proposes to transform it.

3 Radical Realism: The Cultivation of Rational Awareness

Despite the intensive realism-antirealism debate within philosophy of science in the course of the last two decades, there is a widespread conviction within the working scientific community that the best of the bodies of theory produced by scientists can in principle give us knowledge of the independently real. Radical realism is in agreement with that conviction in this limited sense: granted an assurance that in at least some circumstances our rationality is experientially engaged with the real in a way that is not a function of a body of theory, we might then be able to form a persuasive argument that some bodies of theory in some circumstances do provide some knowledge of the independently real. On the other hand, without that assurance it would seem to be impossible to demonstrate that bodies of theory can do this; for in that case we should have to demonstrate from within a body of theory our right to make inferences from the structure of a body of theory to the structure of the independently real. It is true that working scientists lay claim to this right only if they are persuaded that in the long run a given body of theory accords with experience. But what, precisely, is the significance of 'accords with experience'? Many antirealists claim in effect that experience itself is a function of precisely the body of theory that is said to accord with it; and even philosophers of science who are realists will often argue that *any* supposed mode of experience is in fact a function of some body of theory.

In this book let us say that any currently accepted body of theory functions as the *rational pole* of the activity of working scientists, and that whatever mode of experience these scientists consult to determine whether a given body of theory is sound constitutes the *empirical pole* of that activity. Evidently any argument cogent enough to persuade us that a given body of scientific theory affords knowledge of the real (in the qualified sense of the previous paragraph) must first persuade us of the truth of the following propositions: (a) the empirical pole of science is in some measure independent of the rational pole; (b) when we turn to the empirical pole to assess the worth of a theory, we are turning to the real; and (c) this assessment can be effective only if we can know the real when we make the turn. Our direct knowing of it may differ from the kind of knowledge the body of theory that forms the rational pole purports to provide; but without in some way knowing what we find at the empirical pole, how can we even begin to demonstrate that the body of theory does what it purports to do? Realists in the contemporary debate within philosophy of science do not feel the force of this question, and so in their arguments they mill about *within* what I am calling the rational pole, much as they would if they were in fact antirealists; so, in any event, runs the argument of Chapter 3.

Radical realism claims (a) that we have the power of achieving rational awareness of the real, and (b) that we deploy that power whenever we attend *rationally* to whatever particular pattern of entities and events makes up the empirical pole of that body of theory. All such entities and events belong to what we usually call the world of common sense, but that rich and not very precisely defined world is not merely the place where our science begins and ends—not merely the complex of phenomena that, pragmatically speaking, needs to be saved. What we know directly there is an ontic level of great importance, and when rational awareness deploys itself there, it does not do so in the interest of science only. We have already seen that radical realism purports to bring about an actualization of first philosophy, and even that delicate business, which so concerns the relation between our knowing and the real, has its beginnings, if not precisely in common sense, then at least in our attention to the status of common sense. As to that status, it will become clear enough later that the world of common sense is an ontic level in a far deeper sense than that the items we find there belong to the so-called ontology of commonsense language.

We are, of course, already engaged in that delicate business. If we find that we can in principle achieve rational awareness of the real, and if what is available to our attentive glance at the empirical pole of science is in fact the real, then rational awareness of the real is achievable at both the beginning and the end of the theoretic-empirical cycle that constitutes science.[3] To hold out this prospect is to claim that the empirical pole of science, which we ordinarily think of as the *experience* that science consults, should never be thought of as mere experience or as that will-o'-the-wisp pure experience; nor should it be thought of as a formed, or constructed, experience—a by-product of our linguistic power in response to we know not what, as so many contemporary philosophers believe. It should rather be thought of as the cognitively accessible real, accessible in various ways besides the way of science.

A most important way is the reflexive exercise in first philosophy in which we now set about assuring ourselves that the empirical pole of science is a knowable reality not formed by us. If the exercise is successful, the theoretic-empirical cycle of science will then be seen to be *cognitively* accessible as an activity that really does swing between two realities: (a) that of

---

3. In the past I have sometimes called this cycle the speculative-empirical cycle and sometimes the theoretic-empirical cycle. Both names have some merit; in this book it will be convenient to use the latter whenever it seems important to keep in mind the point that the rational pole of science always involves either attention to or production of a body of theory.

well-constructed bodies of theory (real as *constructed* and real as constructed on the basis of reality that is not a body of theory), and (b) the commonsense reality attended to by our rational awareness both before and after the doing of science. The occurrence of various forms of the term 'real' again and again in what I have just written may seem excessive, and so I hasten to assure the reader that it is deliberate rather than a stylistic inadvertence.

Let us anticipate this exercise by making a preliminary effort to attend reflexively to the power of rational awareness. When I say "attend" I insist that we are not setting out to propound a *theory* in which such expressions as 'rational awareness', 'rational-experiential engagement', and 'rational consciousness' figure. We are trying instead to cultivate an attitude of rational attention. Our goal is not the production of a conceptual or linguistic structure whose interpretative virtues we may then debate, but rather the achievement of a reflexive epiphany—an epiphany in the special sense whose currency in the literary world we owe to James Joyce. Prior to the reflexive epiphany—prior as a real function or activity upon which our reflexive intent is to be directed—there is another epiphany, something so commonplace, so taken for granted, that it seems inappropriate to bring so highfalutin a word to bear upon it. The commonplace epiphany is what takes place whenever any particular knower *here* achieves rational awareness of some particular or complex of particulars *there* whose character or structure that act or instance of rational awareness has had no hand in making. The function of rational awareness will indeed have had a hand in achieving or attaining its own telos; but that is to say no more than that the function completes, or actualizes, itself in attaining the independent object. No doubt many subordinate systems will have supported that completion: they subserve the function, and in doing so they vanish (for the knower) into the ontic level we are now calling rational awareness. Their contributing *is* a vanishing; but of course they do not vanish for the physiologist whose task it is to understand *them*.

The realization, or actualization, of rational awareness consists in the cognitive attainment of the thing attended to; that alone is what comes through to us who inhabit—or rather who *are*—that ontic level. The reflexive epiphany we are now concerned to cultivate is indeed about the commonplace one; but it is not radically distinct from the latter. We are now, if this paragraph should succeed in calling attention to what I intend, merely taking the first steps toward cultivating and enhancing the reflexive component that was present from the beginning in the commonplace epiphany. All this is a scandal from the viewpoint of conventional epistemology, which relies in all its occasions upon the commonplace epiphany but recoils from the scandalous simplicity of its commonplace achievement. Conventional epistemology of a more-or-less realist kind consists, on the other hand, of a complex *theory* about knowing that purports to show that, although (of course) we cannot do what is so scandalous, we can at least infer something about reality from some bodies of physical theory if not from others. As for conventional epistemology of a more-or-less antirealist kind, it comprises an equally complex body of theory that purports to demonstrate that none of our bodies of physical theory gives us knowledge of the independently real. In both cases the scandal of a (nontheoretic) rational awareness of the independently real is dismissed in advance.

Rational awareness, like the scientific knowledge that depends upon it, is a mode of knowing that comprises two poles; but, although science swings constantly between its rational and empirical poles, there is no such cycle in rational awareness. Its two poles are integral and inseparable: they cooperate and interpenetrate in each of its instances, and without that union of the two poles rational awareness does not exist. Let us therefore call the two poles of rational awareness the rational and the experiential, rather than the rational and the empirical as in the case of science. But we can express this point equally well by simply calling rational awareness—as I have already done—a rational-experiential engagement with the real; for to be aware is to be experiencing, with all that means in terms of the involvement of our senses and our bodies in general; and to be rationally aware is to be rationally experiencing. Conversely, the exercise of rationality always includes or terminates in awareness, which is to say that it includes or terminates in something experiential. Our rationality is always involved in experience, always integral with experience, even when we deploy it in some highly formal exercise designed to exclude experience insofar as that is possible.

It is important to say again that our warrant for all these assertions is nothing other than rational awareness. Despite my use of a certain amount of philosophical jargon, I am not now propounding theories or making constructs but rather calling attention to something; and calling attention is nothing more than calling rational awareness into play. It is a naturally reflexive function, for some reflexive aspect coexists with its actualization in the reality of whatever it is directed toward. There is no act of rational awareness that does not deploy its reflexive component in at least an incipient way: you the reader, as you attend now to this paragraph, attend also—though no doubt in a peripheral way—to yourself attending. One terminological dodge that helps emphasize the reflexive dimension of rational awareness is to call it rational consciousness, for the original meaning of 'consciousness' carries with it the sense of being aware of something and being aware also that one is aware. Indeed, the term 'consciousness' is so high in reflexive overtones that in post-Cartesian epistemology it is taken for granted that the independent status of the item attended to is problematic: consciousness itself takes center stage, and the item attended to is understood to be *within* consciousness. Whether the item is a tree or a proposition, any reality status it might have independently of our attending to it is then construed as something that needs to be demonstrated on the basis of a chain of inference that is thought to begin in consciousness and to take place there.[4] So even today the term 'consciousness' must be introduced with care; that is why I have preferred the term 'awareness' at the outset. With these radically realistic disclaimers, however, the expressions 'rational awareness' and 'rational consciousness' may be taken to be interchangeable.

Since the term 'intentionality' is much used today, and often by writers who seem unaware that there is a radical difference between the late-medieval *intentio* and the *Intentionalität* of the phenomenologists, I should add that it is counter to the spirit of radical realism to make the term 'intentionality' interchangeable with either 'rational awareness' or 'rational consciousness'. No doubt those medieval writers who distinguished between terms used in first intention (to refer to things) and terms used in second intention (to refer to other terms) took it for granted that someone *intending* that distinction had a mind and thus what was later called a consciousness. But to make that distinction about terms is certainly not to imply that items attended to by the person that intends are inevitably either within consciousness or formed by consciousness. No doubt some of the philosophers who made the distinction might wish to put a term like 'tree' within consciousness, but the distinction did not require them to argue that being conscious of a tree is to be conscious of something that is within consciousness. If some of them did so, it was for other reasons; there is, after all, a powerful late-medieval subjectivism which prefigures that of Descartes. But the powerful and well-developed subjectivism that begins with Descartes does indeed intervene between *intentio* and *Intentionalität*. More important, there is the figure of Kant, whose doctrine of the constitutive, or formative, function exercised by the union of the understanding and sensibility had a powerful influence on phenomenology. That influence made

---

4. I do not mean to suggest that propositions have a reality quite independent of the formative power of our rationality, but merely that, once formed, they have a reality status that is independent of our attending to them. I discuss the formative power of rationality in Chapter 4.

it inevitable that, when Husserl took up Brentano's term of art *Intentionalität* and made it roughly equivalent to 'consciousness', both terms should thenceforth carry overtones of the formative, or constitutive. For phenomenology in general, any item attended to by consciousness/intentionality is thus a function of something constitutive, formative, or meaning giving in the act of attending itself. For radical realism, on the contrary, to be (rationally) conscious of something is by no means to have contributed *by that act of attending* to the structure of the thing attended to. To say so is only to say that the thing attended to is ontologically independent of the act of attending. Propositions and theories are not really exceptions. Although they do indeed owe their form, or at least part of it, to something formative in ourselves, we do not, by virtue of attending to them, contribute yet another form that hides the one we have already endowed them with. I return to this matter in more detail in Chapter 4.

Rational awareness is a very flexible activity. It can focus on all manner of particulars, like the pen in my hand or the page now before the reader. It can attend, moreover, to particulars within particulars, as I attend now, fleetingly, to the black label on my pen even as I still attend to the pen; or as when I attend to one of the letters on the label, without entirely losing sight of either label or pen. But when it attends to any particular, it attends also to many other particulars round about what it is focused on—page, pen, label, and label letter make up only a few of the many numerable particulars I could attend to in this room. This contrast of one and many, where each one is a particular and the many consists precisely of many such particular ones, is characteristic of all acts of rational awareness. But that contrast scarcely gives us an adequate account of what is then going on; nor are the particulars attended to exhausted by enumerating physical objects like those mentioned. I have, for instance, attended to, and asked the reader to attend to, all the words in this essay so far—just recently, for instance, such words as 'rational', 'awareness', 'experiental', and 'empirical'. Set aside the worth of the words, and how they might or might not cohere to interest or exasperate the reader; the point is merely that we are, the reader and I, rationally aware of them, qua words, as several particular items set beside the many particular physical items in my or the reader's near neighborhood. The natural reflexivity mentioned earlier marches along with all of this: whether it should be pen or word, ourselves attending goes along with—is part of the rational awareness of—each of the things we attend to.

We are only at the beginning of the business of calling reflexive attention to our rational awareness. But it is not too early to ask for more detail about what has been characterized so far only by way of the notion of a reflexive epiphany. In what circumstances does this occur? And how can it help us when that circumstance is gone and we are attending to something else? Is there anything in the moment of a reflexive epiphany that can give us some permanent assurance about our cognitive capacities? Let us look again—if that can be done—at this function that is said to have a rational pole and an experiential pole, integral, inseparable, within any of its moments. I was, in the examples, focusing on particular items; my senses and indeed my whole body participated in that focus; and I was rationally and experientially aware both of those items and of that participation. My rational awareness, however, was by no means saturated with a mere *omnium gatherum* of particulars. I was not, nor could I be, *rationally* aware of what was merely many discrete items—many discrete *ones*. Rationality is as incapable of inhabiting a universe consisting of a mere multiplicity of units as such an imagined universe is incapable of existing.

Turning to the hither side of the items attended to, we find that the satisfaction of rational awareness, its sense of reaching a telos, is not achieved in bare particulars: each *one* thus attended to is not just a discrete one among many such. There is a bond that unites them, and it is not an external bond—no mere business of bringing a number of discrete but similar *ones* under one concept or universal term. Each *one* attended to—the pen, the page, the bunch of grapes in the window embrasure—possesses its unity not as something private to it but as one instance of a one that is universal.

What I called earlier a commonplace epiphany is an assenting response to a common, or universal, unity found within any particular in which our rational awareness completes itself and is satisfied. That unity is echoed by a common, or universal, unity within each of the many instances of our rational awareness. But as we persist in our reflexive exercise we see it to be more than an echo: the unity of rational awareness is in fact the same unity that rational awareness assents to in whatever particular it attends to. It is, however, the reflexive epiphany, intensifying the reflexivity incipient in the commonplace one, that reveals this to us; the rational awareness in which we acknowledge the commonplace epiphany is itself no commonplace one except in the sense that it emerges from our reflexive cultivation and enhancement of the latter. While one enjoys, in the ordinary way, a rational awareness of whatever one attends to in some particular situation, a surge in the intensity of the reflexive component present in all rational awareness brings a sudden assurance that the knower—for in this transformatory moment it is no longer just one's particular self that is in question—has cognitively attained something that is quite independent of the cognitive act that attains it. It is an assurance that the knower has not conferred upon the known thing the structure that comes through in the knower's rational awareness of it.

The consequence of its emergence is that the case for rational awareness does not rest on any one instance of rational awareness. The experiential pole persists in each instance, and our satisfaction in what we know directly therefore suffers no loss in particularity; the union of the rational and experiential poles is vital to the integrity of rational awareness. But in the reflexive intensification of rational awareness that leads to the reflexive epiphany, the rational pole is dominant—hence the growing importance, as we proceed with our discussion, of the intimate relation between universality and unity. It is so intimate that the word 'relation' does not quite express what is at issue; that is why it seems appropriate to think of universality and unity not as two distinct factors in direct knowing but rather as one—the U-factor, as I called it earlier. Thus, the unity of the thing known is common, or universal: it transcends each known thing by applying to all of them. So with the unity of any instance of rational awareness: it transcends any such instance by applying to all. For the same reason, the distinction between knower and known becomes qualified as, in reflection, the knower finds that any instance of its direct knowing is characterized by the same U-factor that belongs to whatever known thing it is engaged with. If we approach this complex matter by beginning with universality rather than unity, we find that the universality we have heretofore associated with terms, propositions, concepts, universals, and forms is lodged within, rather than imposed upon, the unity of any particular known thing. And the knower, deploying universality by way of propositional form, now sees that it can do so because universality is lodged within *its* unity. To put the matter another way: we are not concerned with an *imposed* transcendental unity of the known that has its source in an Ego that deserves to be called transcendental because it *imposes* that unity. We are concerned, rather, with a recognition on our part of an independent factor of reality, coupled with a recognition that we attain it in our rational awareness of any particular. The reflexive epiphany's transcendence of any of its moments, and the dependence of this transcendence on a U-factor in every object and every subject, thus become part of the epiphany itself.

This reflexive transcendence leaves our radically realistic conclusions undisturbed by the many acts, moments, or passages of rational awareness in which there is some doubt or error. The U-factor, persisting through our rational-experiential engagement with all particulars, contributes to a complex satisfaction of rational awareness that accompanies and mitigates our consciousness of doubt and error. The complex satisfaction amounts to this: we are entertaining particular realities that are independent of our act of entertaining them and are qualified by universal reality, which is also independent of our entertaining it; we concede that we are not clear about what precisely the particulars are; we hold fast to the general clarity within which the lack of clarity about particulars is suspended. The satisfaction brings with it the conviction that with more careful attention to what is before us we may in the end achieve a more adequate awareness of the particulars. Our rationality, vulnerable and corruptible as it is, has thus, in all its occasions, a restorative access to reality. The bondage of rationality to particulars must nevertheless be acknowledged. Despite the universality and unity that qualify the subject, despite the universality and unity that qualify its object—despite, that is, the U-factor in knower and known—whatever self-possession we achieve as rational beings is won and nurtured from day to day in the particular situations in which we find ourselves.

My emphasis on the importance of the U-factor in the rational pole and on the dominance of the rational pole in the reflexive epiphany thus requires some qualification. Although the U-factor is insistently present in each instance of rational awareness and in each particular that rational awareness is directed upon, the presence in our experiential range of some complex of particulars is essential to all direct knowing. Accordingly, the satisfaction that is the subjective correlative of the actualization of rational awareness in the independence of both particulars and the universal nature of things in which particulars are embedded is always a rational-experiential satisfaction. We must not look for some more arcane and recondite, more hidden and complex justification than what the deployment of rational awareness, and the rational-experiential enjoyment of the real that goes with it, can provide. And that is the scandal of rational awareness: I here and now, you there and in your now, are rationally aware of things distinct from and independent of ourselves, distinct from and independent of the rational awareness that attains them; things having just the particularity they happen to have; things that are nonetheless bound together in a common unity that is not a function of our rational activity.

We may therefore dismiss the mere theory that anything known must be a function of something formative in the function of rational awareness itself—the mere theory that there is something deceptive built into the very tissue of our knowing. Realistic doctrines are sometimes attacked on the grounds that they amount to a "spectator" theory of knowing. Radical realism, at any rate, is nothing of the kind, for the reflexive epiphany carries with it the realization that it is itself an originative achievement. Moreover, although it is the outcome of a deliberate intensification of the commonplace epiphany, one of its consequences is an access of wonder at the originative achievement of the latter as well. The commonplace one, which is so often either taken for granted or repudiated, now appears as the active function it is: though it is no creative coadjutor in the *making* of the reality it attains, it is nothing if not creative in the miracle of attaining it.

I suspect that readers who object to the metaphor of an epiphany are in fact objecting to the thought that we are possessed of a rational awareness that is directed on the world of experience and is self-justifying as well. This seems to give philosophy an empirical function, which most philosophers would say it does not possess, and at the same time to exempt it from the hard work of argumentative justification. Let me set aside the first imagined objection, which I consider in some detail later. It is important just now to notice that the radically realistic claims made on behalf of rational awareness are implicitly made by analytic philosophers on behalf of our capacity to get to grips with any argument, including the kinds of argument deployed to justify the truth, or at any rate the warranted assertibility, of some philosophical thesis. It is true that arguments in general are constructed things and so owe at least part of whatever reality they possess as arguments to the formative power of the rationality of the philosopher who propounds them. But once constructed, they have a structure that is independent of any cognitive response we make to them; indeed we cannot respond adequately to them unless we grasp *that* structure rather than interposing something else by virtue of some formative "cognitive" response. There is, in short, something self-justifying in our assent not just to the whole sweep of an argument but to the smallest step, the most trivial proposition, indeed the most trivial term in the argument.

Philosophers naturally focus on the argument, but we are concerned just now with the capacity that allows them to do so. That capacity is precisely the rational awareness I have been talking about, now considered as directed toward arguments rather than, say, temporospatial entities or events. The metaphor of a commonplace epiphany applies as adequately to our awareness of a sentence in a justificatory argument as it does to our awareness of a tree or of another person. And this conclusion cannot be evaded by transposing the discussion into the key of today's linguistic obsession, for the grasp of language as such, or for that matter the technical attention to the structure of language, rests also on our power of rational awareness. It is irrelevant that the philosophers I imagine would not dream of using the expression 'rational awareness'; it is what they take for granted that I am concerned with just now.

It is by no means irrelevant that, although arguments are in some important sense *entia rationis* (beings of reason), they are also mediated by physical shapes and sounds. But there is a more general point that lies behind that obvious one: although arguments are not precisely empirical items, an argumentative situation is nevertheless an experiential one—one in which particular arguments and their particular parts are experientially encountered. I do not mean just that physical symbols for arguments are encountered, but rather that the arguments as such are encountered in a situation in which the discrimination of physical symbol from argument is itself an experiential activity. There are, of course, many philosophical doctrines that do not have in their "ontology" any function that entertains, is conscious of, or is aware of entities of any kind. Usually doctrines of this kind dismiss counterpart entities as well, and in this dismissal it is not just *entia rationis* like arguments and concepts that go: even their physical symbols, thought of as entities really encountered, are excluded from the "ontology." These doctrines hold that, when we deal with an argument, we are merely behaving in response to certain linguistic stimuli, and that one way of behaving is to postulate functions like awareness, items like concepts and arguments, and physical things like pages with ink on them. In more extreme versions, it is not rigorous to talk of *linguistic* stimuli, for the real stimuli are thought to be not knowable as such; out of our response to them—understood as a kind of primordial theory holding—we construct commonsense stimuli like sentences and arguments.

Doctrines like those just sketched amount to an attack on the ontic level of the person. Questions about the real relation between philosophers and their arguments are dismissed by dismissing that level as inauthentic. A common device is to make the ontic level of the person equivalent to the commonsense world and then to dismiss the latter either as a product of an inadequate prototheory in the reductionist way just sketched or (in a more "humane" spirit) as one of the many worlds we produce by virtue of our linguistic-cum-theoretic power of synthesis. Radical realism, on the other hand, acknowledges that it derives all its authority from the ontic level of the person. But all construction of philosophical doctrines and all criticism of them—indeed, all argument whatsoever—also derives its authority from there. None of the dodges we have just glanced at—we shall be looking at some of them in detail in Chapters 3 and 4—is free of the fundamental dishonesty of relying on rational awareness in every step of a philosophical maneuver that is designed to dismiss the reality of the ontic level of the person.

4 Rational Awareness of the Relation between Reality and the Propositional

All this introductory material is placed before the reader and, in some measure, before the writer as well by the production of language. From one point of view, the writer has done no more than produce certain sentences of a language that, in two senses, is a received one. His command of English is received and then developed in accordance with his own powers; and many of the characteristically philosophical turns of speech are also received from a going tradition, even if they too are developed in accordance with the writer's own needs, interests, and convictions. No doubt some of these turns of speech are aberrant in the sense that they deviate from orthodox philosophical usage; but they all begin in a received context, and the possibilities of deviation are limited to some degree by that context. I have, after all, chosen to write about realism.

When, therefore, I claim to be calling attention to some particular things, evoking nothing more than a (radically realistic) power of rational awareness exercised by both the reader and myself, am I not in fact merely calling upon the reader to share with me a language I have received and altered in a more or less aberrant direction? Is not what purports to be a self-justifying rational awareness of the independently real in fact merely a function of language? And, if this is so, how can we circumvent a constraint that makes mere constructed "realities" emerge out of the collision of the formative power of language with whatever is not language? How can we get out from under the net of language and be in a position to announce that it is not really a net that inevitably constrains us?

From the point of view of radical realism, it is important to notice that all these objections are cast in the form of a body of philosophical theory that has dominated academic philosophy in the English-speaking world for much of this century. (It is no less a body of theory for being in part about the status of theory in general.) Just here it is worth anticipating two points from the more extended discussion in the next two chapters: (a) this body of theory conflates the notion of language with the notion of theory, with the consequence that the relation between language and reality tends to be confused with the relation between theory and what theory purports to illuminate; and (b) it also claims that what theory illuminates is not accessible by any approach that is not linguistic-cum-theoretic. But if that body of *philosophical* theory is not sound, then the real relation between our natural language and reality may be quite different from what it says or implies. One obvious difference is that we shall no longer have to conflate language and theory. No doubt the relation between language and reality is a difficult one to express, but the fact that we must express it in a language offers no special difficulties. Certainly we need not be put off by the thought that the effort is so logically peculiar that we are doomed in advance to failure: that would be true only if the body of philosophical theory accepted by the linguistic consensus were sound.

It follows from all this that expressing something is not inevitably propounding a body of theory about it—still less propounding a body of theory-cum-language about it. There are cases where nothing but a body of theory can help us: we cannot even say what we mean by the claim that there are quarks, let alone decide such questions as whether quarks are more fundamental than electrons, without turning our attention to a body of theory so difficult that most of us must see it as in a glass, darkly. Language, however, together with some features of its relation to nonlinguistic reality, is an accessible enough thing—which is to say, within the setting of the reflexive exercise that is the beginning of radical realism, that it is something of which we can be rationally aware, distinguishing it as we do so from other things of which we can also be rationally aware. In this rational awareness of two distinct things—nonlinguistic reality and the reality of language itself—their relation also comes within the scope of rational awareness. Some of the functions of language also come into view: thus, I am now trying to call attention to certain things and aspects of things, and I am using language to do so; but I am also calling attention, reflexive attention, to the fact that one function of language is the calling of attention to what is nonlinguistic.

One oddity about language is that its coming into being introduces a doubleness into our consciousness. Invent a name for one particular thing or for several resembling things—surely an innovation that must be close to the beginnings of language—and you have introduced the name, as an at least incipient object of attention, alongside whatever it names; and this doubleness probably long preceded the invention of the name 'name'. To say so, to be sure, is to propound a minitheory concerning the origins of language, but the remark does rest on what is in no sense a theory: our *present* ability to discriminate between a name attended to and a thing attended to. With this ability goes a more complex one we have long possessed: the ability to attend either to a proposition or to the real situation it expresses and then to distinguish those two things. Our rational awareness of the proposition 'snow is white', qua proposition, is quite a different matter from our rational awareness of snow on the ground. One extreme form of the body of theory about language which I have temporarily set aside would have it that 'rational awareness' is a superfluous notion; that whether we are attending to proposition or thing, what is in fact going on is a language processing; that language processing *is* what I have been calling rational awareness; and that this processing takes place in response to stimuli, the stimuli themselves not really being objects of rational or any other kind of attention. But, setting such theories aside, the distinction seems real enough.

I therefore appeal once more to the commonplace epiphany, itself attended to by its heightened reflexive counterpart: we can distinguish between the nonlinguistic and the linguistic as objects of attention; we can attend to the proposition emerging from the nonpropositional—emerging from it in the sense that it is a consequence of our attending to the nonpropositional rather than the other way round. As we make this reflexive move, we turn our attention to a supposedly inexpressible matrix out of which, according to a common contemporary dogma, "reality"[5] springs as a function of language-cum-theory. Something does in fact spring from there: first, the reality of language itself as a (constructed) syntactic structure; second, the reality of our real experience reconstructed in its absence in our consciousness, or of imagined experience constructed in our consciousness out of real experience; third, drawing on both first and second, constructed bodies of theory involving both postulated entities and aspects of real entities. In all these cases we are dealing with realities that are authentic constructed realities—that is, realities that are indeed constructs and are moreover recognized to be just that. They are by no means Pickwickian realities, because they do not purport to be what they are not.

Language, considered as a constructed reality, is but one feature of an experiential situation—*some* experiential situation—which we recognize as nonlinguistic by virtue of the same rational awareness we use to focus on language. We cannot either use language or recognize it for what it is except against the background of some nonlinguistic situation of which we are also rationally aware. Whatever else language is, it is also sounds and shapes, which we can understand and use only if we can already discriminate sounds and shapes. Once we cease our obsessive attention to a certain body of philosophical theory and allow rational awareness to have its way, a common post-Kantian epistemological confusion—the notion that before acquiring language we live in a booming buzzing confusion—loses its obsessive force. The baby being taught to speak discriminates speech sounds from other sounds and links the former with discriminated shapes, movements, and actions. Someone whose first sentence is "Turn the light on" has long since discriminated the light, the action of turning the switch, and the connection between that

---

5. This is reality in the Pickwickian sense invented by certain antirealists to persuade us that all supposed reality is in fact an outcome of a formative or constructive power inherent in our language use; see Chapter 4.

action and the appearance of the light; that child, moreover, has already discriminated the sounds that are words from other sounds.

No doubt we create various "worlds" with the help of language, but it seems safe enough to say that, whether these "worlds" belong to art, to theory, or merely to reverie, the imagination also plays a role in their making. The formative power that makes such "worlds" is complex, and it is difficult to isolate the contribution of language from that of other factors. It seems clear enough—setting aside the blandishments of the philosophical body of theory I have touched on—that by virtue of language we can call attention to items, activities, and aspects of a nonlinguistic world; that we can draw on our memory of such things with the help of language; and that a formative power that includes but is by no means limited to what is usually called imagination can respond to all this by producing all manner of beautiful and useful things. The metaphor 'world'—and it should not be forgotten that it is indeed a metaphor—can no doubt be applied to many of these things. It is striking and in some ways illuminating to speak of the world of Dante, or the world, or worlds, of Star Trek, or the world of Newton's *Principia,* or the possible worlds considered by modal logicians and their followers in theology and other fields. But they are all "worlds," and they are all within the world, even though some of them may provide us with knowledge about the world that could not be had otherwise. But the business of the making of such "worlds" gives us no reason whatsoever to suppose that the world, in which the activities that make use of the metaphor have their roots, is itself the product of a rational formative activity on our part—still less of a merely linguistic formative activity.

In general, there is a distinction between whatever formative function rationality possesses and its function of rational awareness. We shall, however, be in a better position to consider the distinction between those two basic functions of our rationality after the next two chapters, in which we shall also be examining in more detail the philosophical doctrines that have done so much to obscure that distinction.

5 Dismissing the Foundation Metaphor

Many writers of the modern era who thought of themselves as pursuing what I am here calling first philosophy have made use, in one way or another, of Descartes's metaphor of a foundation for knowledge. Because the real, with which, according to radical realism, we are cognitively engaged, is what those writers were looking for, radical realism makes common cause with at least their intent, if not their conclusions. The reader may therefore wonder whether I think of radical realism as providing a foundation for knowledge, the more so because, in at least two publications and in certain unpublished preparatory studies delivered as lectures, I have tried to breathe new life into a notion that most philosophers have now given up on.[6]

There has been, I think, much confusion about the notion of a foundation for knowledge. There is one foundational tradition that is identical with the kind of metaphysics I wish to call first philosophy; there is another that is antimetaphysical—its beginnings coincide roughly with the beginnings of empiricism, and its end came about the middle of this century; and there is also a doctrine called foundationalism. The term 'foundationalism' seems to be a relatively new one. The doctrine it names is usually an attempt to revive the antimetaphysical foundational tradition; more rarely it is an attempt to revive the metaphysical foundational tradition. If most philosophers have given up on the notion of a foundation for knowledge, it is because foundationalism has generally been judged to be a faulty doctrine.

---

6. Edward Pols, *The Acts of Our Being: A Reflection on Agency and Responsibility* (Amherst: University of Massachusetts Press, 1982); "After the Linguistic Consensus: The Real Foundation Question," *Review of Metaphysics* 40 (September 1986): 17–40.

The foundational tradition that is concerned with perfecting a first philosophy manifests two conflicting themes. One theme—it is in fact much older than the metaphor, but it is at least implicit in the work of many who use the metaphor—suggests that the real foundation never consists in a set of principles expressed in propositions. Sometimes, especially when the influence of neo-Platonism is felt, it is also suggested that there is something about the real foundation that is not expressible in propositions; sometimes, especially in idealism, merely that the real foundation is to be found in an active power capable of expressing things in propositions. The other theme is that a foundation for knowledge must consist of a set of principles, that is, a set of *fundamental* propositions. This latter theme can be found in both the foundational tradition that affirms first philosophy and the one that rejects it; it is especially salient in what has come to be called foundationalism.

It should be clear enough that radical realism wishes to call attention to a rational function or activity that is not static in the way propositions are static, however bound up with the development of language the function may be. And it is this static character of propositions that goes best with the static nature of the foundation metaphor itself. According to radical realism, the origins of the propositional are to be found in the overlap of the active function of rational awareness with the equally active formative function of rationality. In my own efforts to revive the foundation metaphor I tried to focus on the activity that *makes* a foundation; I sometimes used the expression 'foundation-directed exercise' to draw attention to it. But I now think that the static overtones of the metaphor itself make that a self-defeating maneuver. What we need instead is the authority of an *activity*—an activity that is dynamic and generative of the propositions that partially express it and sure of itself in the same active and self-confident mode that gives rise to propositions. If radical realism is in due course persuasive, the foundation metaphor will lose what hold it still has and be displaced by such expressions as 'rational authority', 'rational autonomy', 'self-confident rational activity', and 'reason-in-act'. I call them 'expressions' because I do not think they will be functioning as new metaphors that displace the old one.

If we consider not the systematic outcome of so many foundational enterprises but rather the effect of the metaphor on the rational powers of the philosophers who undertook them, it seems fair to say that the metaphor too often represses certain resources of our rationality whose free play is essential to our seeing clearly and satisfying the persistent need for reality that is so central to first philosophy. That the metaphor was invented precisely to help satisfy that need does not alter the matter. In the present setting, then, the most important objection to the metaphor is that it fails to do justice to the active nature of the power or powers by virtue of which we know and by virtue of which we produce the complex propositional structures in which knowing is expressed, stabilized, and made communicable. The foundation metaphor—with all those overtones of something that supports because it is static, inactive, and enduring; of something that does not actualize itself in the temporal order—strikes quite the wrong note. It also helps encourage that most hubristic of all aspirations of first philosophy and science, the aspiration to an ideal, complete, and perfect knowledge consisting of a set of propositions bound together in a deductive unity. What we need instead is a confident self-knowledge that is coincidental with our active dealing with the independently real, one that does indeed sustain the judgments we make about the propositional products of our rational activity but whose autonomous vitality is not to be found in those products as such.

This brings us back to the question of the relation between reality and the propositional. The point of all this emphasis on activity is by no means that reality is inexpressible. It is rather that what we do succeed in expressing in propositions has precisely the status of a formed and completed group of *propositions*. If we then take them as our foundation, we have overlooked, put out of play, or dismissed the rational activity which, in its experiential engagement with the real, formed or produced those same propositions. That activity is never completed; indeed, it is a permanent background activity sustaining those propositions all the while it is itself sustained by what it is engaged with. It sustains those propositions in the straightforward sense that certain marks or sounds become vehicles for propositions, rather than mere sounds or marks, only in the presence of rational action. Language does not make reality. Language is made by rationality—itself a nonlinguistic reality—when rationality so engages the real as to become aware of some part or feature of it. So our authority lies in the activity of rational awareness in its engagement with the real, an activity that involves language making and language sustaining but is by no means identical with them. And these propositions I now enunciate have as their deepest function the calling attention to something in every respect translinguistic.

The corruption of language consists in our substituting it for the real: we seduce ourselves into supposing that something made by the same formative function that makes language—something that is certainly in part linguistic but also in part something imagined—is in fact the real. But I do not mean merely that bodies of false theory are often confused with the real things we are interested in. That, of course, happens more than we like to think. Thus, much of the twentieth century had been dominated by such bodies of theory as human-nature-as-imagined-and-articulated-by-Freud, history-as-imagined-and-articulated-by-Marx, the function-of-language-as-imagined-and-articulated-by-Wittgenstein, human-culture-as-imagined-and-articulated-by-Foucault. In contemplating such bodies of theory, myriads of intelligent human beings have supposed that they were in fact contemplating the realities their interest in which had led them to turn to those bodies of theory in the first place. I mean rather the further seduction of supposing that such formed "realities"—which are, to be sure, in part formed on the basis of reality—are the only things accessible to us; that in fact the distinction between Pickwickian reality and reality is a false one.

3

• The Linguistic Consensus

1 *The Conflation of Language and Theory*

• For the greater part of this century, a certain dogmatic complex has so dominated analytic-empirical philosophers that it has prevented them from noticing that the reality question, for all its difficulty and subtlety, calls for a directness and simplicity on the part of the inquirer that is radically different from what professional philosophers are used to. In the previous chapter I identified an appropriate route of inquiry with a nonspeculative, nontheoretic first philosophy that is so far only an aspiration. We shall return to that route in Chapter 5. In this chapter and the next, I deal with the dogmatic complex that has done so much to confuse the issue—confuse it not just for those who have propounded the dogmatic complex and are still in its grip but for our culture in general.

Early in this century, analytic-empirical philosophy began to make language the focus of its interest. Surveying those times, our retrospective glance reveals two strands already present in that interest, for the term 'language' was applied not only to natural language but to formal systems as well, especially those designed to be interpreted in terms of the principles of logic and, by way of that, in terms of the principles of mathematics. Indeed, the expression 'natural language' owes its currency to the need to distinguish a variety of formal languages from the background language out of which they develop and within which they have the only life they do have. For reasons that need not detain us here, I take only natural language to be language in the strict sense and therefore regard other uses of the term 'language' as metaphorical extensions of that home sense. Leibniz's powerful use of the metaphor is there to remind us that that extension is no twentieth-century invention; but analytic philosophers have made such generous and systematic use of it that it is fair to say they have conflated the notion of language we derive from our familiarity with natural language first with the notion of a formal system and then with some other notions that are of great importance for natural science—especially the notions of hypothesis and theory.

This conflation has produced two related assumptions that have been powerful within the analytic community: that there is something defective about natural language; and that it is the task of an enlightened intellectual community to replace natural language with a canonical, perfect, or ideal language. A less noticed but influential consequence of the conflation of the notion of language with the notions of hypothesis and theory (in the narrow sense), and with the wider notion of a scientific body of theory that includes hypotheses, theories (narrow sense), assumptions, laws, models, and a great deal more besides, is an erosion of the notion of language itself. What had begun as a kind of imperialistic encroachment of the term 'language' on other spheres theretofore distinct is transformed into an imperialism on the part of such terms as 'hypothesis' and 'theory'. In the end what purports to be natural language—the very source of our metaphor—is now understood by some philosophers, especially realist philosophers of science, to be a complex set of hypotheses and theories; and by and large these are alleged to be mistaken ones, and so in principle replaceable without significant loss. One might well regard this conflation of the term 'language' and other terms that are, in a prima facie sense, distinct as one of the dogmas that go to make up the linguistic consensus, but the conflation has been so much a matter of a gradual but relentless tendency that it will be convenient to think of it as a background for the dogmas themselves. Except for this development, the imperialism of the notion of language is real enough; we notice it especially in the influential doctrine that the only rigorous way to talk about what is or is not real is to talk of what does or does not belong to the ontology of whatever language we are concerned with. This sense of the term 'ontology' is now so common that it seems to have been quite forgotten what the term originally meant.

2 Seven Dogmas of the Linguistic Consensus

The dogmatic complex that concerns us pervades the work of a large group of philosophers, most of whom belong to the English-speaking world and consider their own work to be analytic. They are a diverse group; the realism-antirealism controversy is only one of many issues that divide them. The dogmas that bind them together, however, make them, when all is said and done, one of the most powerful and unified intellectual communities on the contemporary scene. It is not in essence a political community, nor do the dogmas that define it have any immediate political consequences. There are, however, other contemporary movements that share at least some of the views about language that define the consensus, and the confluence of one of these movements with the one that concerns us has had political consequences that we should glance at before turning to the dogmas themselves.

The movement in question is structuralism, in which I include so-called post-structuralism—in my view, better called late structuralism. Structuralism has some of the same philosophical roots as analytic philosophy, but also some rather different roots in linguistics and anthropology. It has, moreover, never shared the analytic community's interest in philosophy of science of the kind that is oriented toward physics. These differences, together with its more speculative and rhetorical style, set structuralism somewhat apart from the linguistic consensus as defined in this chapter. But there has indeed been a confluence of the two movements, so much so that it is fair to say that they form a wider and more powerful linguistic consensus than the one we shall be concerned with. The later structuralists—especially those who have been in the forefront of the deconstructionist movement—have made common cause with certain relativistic tendencies of the linguistic consensus, drawn on their own profounder relativism to focus and magnify those tendencies, and codified them in a doctrine that is of great utility in political activism. It is the doctrine that the pursuit of power, rather than truth, is not just a common failing of philosophers, as most of us have supposed it to be, but rather the very essence of philosophy itself.

It is essentially this doctrine, generalized to a doctrine about all nonscientific disciplines, that has furnished a kind of intellectual justification for those who wish to bring their political convictions in national and world politics—more particularly, their advocacy of certain interest groups—to bear upon the affairs of the academy. In this book we shall not be directly concerned with the politics of either the academy or the world outside it. Nevertheless, an examination of the dogmas of the linguistic consensus (in the narrower sense that concerns us in this chapter) should make it clear why a politically useful doctrine like that should inevitably emerge from the cultural context to which the consensus has been so notable a contributor.

Some parts of the dogmatic complex we now turn to are quite explicit—as explicit as items in the creed of this or that religion—and like such items they are no less dogmatic for being open to sectarian interpretation. Other parts of it are far from explicit and so must be inferred either from the declared goals of writers or from what they refrain from discussing. Like all dogmas, some may be true, some false; the fact that they are dogmas does not help us decide. What is essential to their being dogmas is that, for those under their sway, there are no alternatives that can be respectably entertained.

There are no doubt many ways of unfolding the dogmatic complex. I will unfold it into seven related dogmas, and I will say that writers who accept the complex belong to the linguistic consensus, which I will sometimes call instead the dogmatic consensus. The complex provides a framework within which disputes that appear to be fundamental can go on without calling the framework itself into question. It provides the setting for the recent debates about realism and antirealism in philosophy of science and so remains undisturbed by their outcome. Brought together within its familial embrace, Hartry Field and Hilary Putnam are no less philosophic brothers than they once were. Within this philosophic college, Peter Strawson and Michael Dummett can dine amiably together at high table, and both will find conversable more than one Antipodean realist who should appear there as a guest.[1] And W. V. O. Quine, Wilfrid Sellars, Nelson Goodman, Thomas Kuhn, and Richard Rorty, schooled in a regimental mess in which the one thing not done is to deny that the knowable is the linguistic, will not cut a messmate who inclines, in some circumstances, to a genial linguistic relativism; or another whose commitment to some one language is so strong as to countenance no word but that momentous one 'reality' to characterize its cognitive grasp; or yet another—Rorty is a case in point—whose pragmatism is so generous as to seem to embrace both of these attitudes.

I. THE DOGMA OF THE CRITERION FOR KNOWLEDGE. Knowledge is a matter of justified true belief; alternatively, it is a matter of correct, or warranted, assertibility.

---

1. For instance, Robert Nola and Michael Devitt. For the former, see " 'Paradigms Lost, or the World Regained'—an Excursion into Realism and Idealism in Science," *Synthese* 45 (November 1980): 319–50. For the latter, see "Dummett's Anti-Realism," *Journal of Philosophy* 80 (February 1983): 73–79.

It would be conveniently simple if all members of the linguistic consensus who are realists settled on the criterion of justified true belief and all who are antirealists (nonrealists) opted for Dewey's pragmatic criterion of warranted assertibility. It seems, however, that as time goes on the coherentist principles of antirealists tend to dominate even those consensus philosophers who once inclined to scientific realism in a downright, old-fashioned sense that seems to cry out for nothing less than the criterion of truth. The sixth dogma and the discussion of the realism-antirealism issue in Chapter 4 will perhaps make it clearer why this should be so. But to be comprehensive we should perhaps include both criteria.

II. THE DOGMA OF THE PRIMACY OF PROPOSITIONS. Knowledge consists of a system of propositions, and it is therefore these in which the knower has justified true belief; alternatively, it is these that are correctly assertible.

The term 'proposition' must therefore cover a good deal, for clearly a body of knowledge may comprise theories, doctrines, hypotheses, models, formal systems, mathematical objects in general, and (formulations of) laws. And this catalogue is by no means complete: bodies of knowledge as extensive as that of twentieth-century physics comprise many other items as well. But it accords with analytic-linguistic practice to classify most items of this kind as complex propositional structures, and all such structures are themselves propositions under one very general acceptation of the term 'proposition'. The term 'proposition' itself, although widely used, may be replaced by 'statement' or 'sentence', for by no means all those who adhere to the dogmatic complex acknowledge an ontological commitment to entities called propositions.

For the consensus, empirical knowledge is the exemplar of knowledge in the sense that even the most formal of intellectual pursuits are most esteemed because of their eventual contribution to natural science. Nor are there any difficulties for the consensus about identifying a body of propositions that constitutes empirical knowledge. But in general it is not in the spirit of the consensus to suppose that the identification of such a propositional complex is something philosophers are professionally competent to do. The determination that a body of propositions does indeed qualify as empirical knowledge is thought instead to be the prerogative of the community of scientists.

III. THE DOGMA OF THE COGNITIVE EXEMPLAR. The scientific community provides the exemplar of knowledge. Knowledge is therefore, in an important sense, a social thing, for it is the body of scientific propositions actually accepted by the scientific community that constitutes exemplary knowledge. On the other hand, that body of propositions is worthy of acceptance because it is empirically sound—that is, because it takes account of that which is in some fundamental, but by no means clear, sense nonpropositional.

It is no longer a tenet of the consensus that physics is the only discipline that meets the most rigorous standards for science, or even that natural science is the only body of knowledge. Nevertheless, science is the only body of propositions accepted as knowledge by the consensus, and physics is the only science that all members of the consensus take to be truly science. In any case, the dogmatic consensus takes it for granted that there is a body of empirically sound knowledge, and it also assumes that scientists know how to determine when a body of propositions is indeed empirically sound (does indeed take account of the nonpropositional). There is, however, little agreement about how that empirical soundness should be interpreted by philosophers—hence the realism-antirealism controversy in philosophy of science. The neutral term 'nonpropositional' allows us to remain uncommitted about the use of the term 'real', as we must do if we are to be faithful to the present state of the dogmatic consensus.

The fourth dogma concerns what the discipline of philosophy itself can do and ought to do and what it cannot do and ought not try to do. It is a subtle matter, and it is not prudent to try to compress it into a brief formula.

IV. THE DOGMA OF THE EMPIRICAL IMPOTENCE OF PHILOSOPHY. It is the job of philosophy to provide a rigorous account of the conditions under which a body of propositions wins acceptance by the scientific community as knowledge of the nonpropositional something-or-other that is its empirical ground. Such an account, it is understood, is itself a body of propositions—propositions that deal not with the empirical but rather with such topics as logic and formal systems, reference, meaning, justification, truth, and assertibility.

Thus, although this philosophic account must also qualify as knowledge in the sense laid down by the first two dogmas, it does not do so by exemplifying the relation between propositions and the nonpropositional which exists in the case of accepted science—whatever that relation may be. Its job is to give whatever elucidation of that relation can be given, not to exemplify or prescribe for it. But the propositions of philosophy are not about the nonpropositional, nor even about some supposed empirically available relation between propositions and the nonpropositional, but rather about the propositions that compose an accepted body of knowledge. Whether the propositions of philosophy bear an *appropriate* relation to an accepted body of scientific knowledge depends upon their meeting certain standards of logic, analysis, and argument; philosophy has no empirical function.

The next three dogmas touch on the concerns of traditional epistemology. The fifth dismisses that kind of epistemology of which the train of thought running from Descartes through the other Continental rationalists, the British empiricists, Kant, and the nineteenth-century idealists is our model. That traditional epistemology purported to do what most members of the consensus try to do with semantics or, here and there, with hermeneutics; but traditional epistemology had other ambitions as well. It included a variety of attempts to focus on the knower, considered as subject, and on the structure of the cognitive act. There is a restricted sense—usually overlooked in the setting of twentieth-century empiricism—in which much of that epistemology can reasonably be said to be empirical in intent, for the reflexive turn so characteristic of all of it includes an effort to attend experientially to the knower at work. My formulation of the fifth dogma is one way of expressing a very general disposition on the part of the consensus to exclude the findings of such a reflexive turn from any rigorous account of knowledge. Any supposed subject matter that requires the use of such terms as 'subjectivity', 'consciousness', 'ideas', 'mind', 'mentality', and 'intentionality' is therefore regarded by many and perhaps most members of the consensus as highly questionable. Those consensus figures who countenance some of these terms usually go to some lengths to assure their readers that they are doing so in a qualified and language-bound sense. Some antirealist members of the consensus who in their more recent views tend toward idealism or at least toward phenomenology—which in the hands of its most influential advocate, Husserl, also tends toward idealism—have now taken up the term 'intentionality' and sometimes also the term 'mind', used in a sense controlled by the sense they give 'intentionality'. But this concern with intentionality does not carry with it a call to the knowing subject to attend in a reflexive and experientially-grounded way to its own cognitive acts. Intentionality is regarded instead as either a feature of certain kinds of terms and propositions or as a meaning-giving function inferred from the meaning found in language.[2]

2. It is true, however, that the term 'intentionality' was deliberately

The fifth dogma, then, may be understood as an extension of the fourth dogma to the special case of our recurrent attempts to focus on ourselves as knowers. The formulation of it must allow for the fact that the consensus comprises some who are reductionists and some who are not.

V. THE DOGMA OF THE PHILOSOPHIC IRRELEVANCE OF THE KNOWER'S REFLEXIVE ACCOUNT OF THE COGNITIVE ACT; OR, THE DOGMA OF THE IRRELEVANCE OF THE SUBJECT. An effective philosophic account of knowledge can and should be drawn up without including an account of whatever activity one supposes one exercises or enjoys when one entertains an article or a body of knowledge. The prohibition applies not only to the case in which one entertains a mere proposition or mere propositions but also to the case in which one entertains something, by way of sensation, that purports (before philosophy intervenes) both to be nonpropositional and to authorize the assertion of a proposition or propositions. It is presumably impossible for the knower to attend reflexively and rationally to some activity, function, state, condition, or property of its own that is the counterpart of what is known, even if it should be shown that such a counterpart had a distinguishable ontological status. In any event, such an enterprise would be philosophically irrelevant.

---

given a resonance with such terms as 'consciousness' when it was invented in the early stages of the phenomenological movement. I say invented, rather than revived, because *Intentionalität,* as introduced by Husserl in response to Brentano's revival of the medieval notion of intentional (in)existence, was by no means a simple revival of the medieval doctrine nor even just a recasting of that doctrine with a Cartesian twist. In the hands of the founder of modern phenomenology, the term implied both a meaning-giving, or constitutive, function and the involvement of that function with consciousness. The term *Intentionalität* has thus a post-Cartesian resonance (because 'consciousness' is first used after Descartes, although perhaps partly in response to his "New Way of Ideas") and a Kantian and neo-Kantian resonance as well (because intentionality is taken to be constitutive). See also the discussion of intentionality in Chapter 2, Section 3.

It should not be thought that the consensus merely rejects the doctrine that we have a more or less Cartesian consciousness or subjectivity that is capable of attending to itself. For if one were to claim, as I have already done in a preliminary way, that to know is to exercise a rational awareness that is in the first instance a sense-based way of attending, and that that way of attending can be turned to profitable reflexive employment without losing its experiential footing, the consensus would reject that claim too. From that perspective the fifth dogma might well be called the dogma of the irrelevance of rational awareness.

But philosophers who insist that we cannot attend reflexively to the activity by virtue of which the knower entertains the known may nevertheless have some definite views about how that activity operates in certain circumstances. And in fact they do have a long-standing set of views on how that activity engages, by way of sensation, whatever purports (before philosophy intervenes) to be the nonpropositional, or empirical, ground for the warranted assertibility of propositions. Those views so dominate the consensus that no member has tried to make them fully explicit. The whole issue is so subtle that it is advisable to explore it carefully before giving it summary form in the sixth and seventh dogmas.

The consensus supposes that the activity by virtue of which the knower empirically engages the nonpropositional, and so is warranted in asserting a relevant set of propositions, cannot appropriately be described as the recognizing, finding, discovering, attending to without altering, or entertaining without altering of something whose nature, *as known*, is independent of the activity of the knower in the sense that it is not the consequence or outcome of that activity. Consensus *realists* do, however, suppose that a certain *body of theory* produced by the knower can in principle provide knowledge of the independently real; they take it for granted that this supposition, which I consider in the next chapter, is consistent with the negative supposition I have just attributed to the entire consensus.

This negative supposition has a positive counterpart—that this same activity of the knower is a formative, constructive, or productive one but not a radically creative one. The knower, that is to say, is thought to produce what is known but to do so only on the basis of something it does not produce. Whether the outcome of the activity is a mere proposition (or mere propositions) or something that is propositional without consisting only of mere propositions, the activity cannot come into play unless the knower is supplied with something nonpropositional. The nonpropositional may therefore be said to be a causal factor in the production both of mere propositions and of that structured experience which, although it is not an experience of mere propositions, is—so at least they claim—nonetheless propositional. Alternatively, the nonpropositional is the matter for which rationality supplies the propositional form.

Despite all the talk within the consensus about the abandonment of the illusion of the empirically given, there is therefore a sense of givenness—different to be sure from the one the consensus once held but has now abandoned—that rescues the consensus from a radical idealism, despite the idealistic overtones that have been noted here and there within it. For the nonpropositional must indeed be given (sense 1), that is,

> given as material for a construction,
> given for worldmaking,
> given to be known by way of the imposition of propositional form,
> given to stimulate the production of paradigms it is nevertheless independent of,
> given to be a determining factor in theory which it nevertheless underdetermines,
> given to be pragmatically coped with.

But in the sense usual in discussions within the analytic-linguistic consensus (sense 2), it is not given—it is not, that is, given so as to be received *as given* by the activity, state, or condition of the knower which is the counterpart of warranted assertibility.

There is thus a distinction between the given (sense 1) and what rational awareness declares it to be or (mistakenly) supposes that it *finds* it to be. Though given (sense 1), it cannot be taken as given, and so it is not really given (sense 2).[3]

VI. THE DOGMA OF FORMATIVE RATIONALITY; OR, THE DOGMA OF LINGUISTIC ENCLOSURE. The items or entities entertained by rationality when it knows, in the sense proper to the first and second dogmas, are literally—and merely—propositions. But what rationality entertains is propositional (or linguistic) not just when rationality is directed on a complex of propositions as such (say a body of theory) or on a single proposition but also when (before philosophy intervenes) it supposes itself to be directed, by way of the senses, on concrete items or entities whose natures, *as entertained,* are not propositional. All entertainment whatsoever on the part of rationality is the outcome of its own formative, constructive, or productive activity. And the nature of this formative activity is best conveyed by saying that the outcome of it is propositional (or linguistic) even when what is entertained is

---

3. Discussion of the given has been mainly in a setting that descends from the failure of the positivist attempts both to find empirical certainties (without commitment to the existence of commonsense entities) and then to pin them down in propositional form; see the discussion of Schlick and Neurath in Chapter 4, Section 1, especially note 2. But it should be remembered that that whole effort was predicated on the assumption that not even commonsense entities, let alone entities of other kinds invoked by metaphysicians, could be given (sense 2) to rational awareness. The positivists' effort was a last-ditch one to respond rationally to a supposed sensory given (sense 2). But it was already taken for granted in positivist circles that the constructive activity that undoubtedly goes into theories was canonical for any cognitive response rationality can make to whatever affects it, so commonsense entities were understood to be constructions out of the empirically given (sense 2).

not merely a proposition or propositions but purports (before philosophy intervenes) to have a nature which, qua entertained, is nonpropositional. Alternatively, rational *experience* is linguistic.

Therefore, if there were indeed items, entities, and even a whole world having natures not dependent upon this formative activity, there would be no sense in which those natures could be entertained as *those* natures by rational awareness.

In Chapter 4 we shall be considering the significant differences between consensus realists and antirealists, and it is therefore important to notice that, although the role assigned to the formative activity in the sixth dogma suggests that only antirealists could accept it, it is in point of fact accepted by consensus realists as well. The only realism open to members of the consensus is one that accepts an overarching antirealistic constraint. The sixth dogma lays it down that all efforts "to break out of discourse to an *archē* beyond discourse" are fruitless.[4] Knowledge is a relation to propositions, and justification consists in "propositions-brought-forward-in-defense-of-other-propositions"—a regress that is at least potentially infinite.[5] Not all members of the consensus would agree with Rorty that justification is a social phenomenon, but I think that all would agree with him when he goes on to say that it is not "a transaction between 'the knowing subject' and 'reality'."[6]

4. Wilfrid Sellars, *Science, Perception and Reality* (New York: Humanities Press, 1963), p. 196.
5. Richard Rorty, *Philosophy and the Mirror of Nature* (Princeton: Princeton University Press, 1979), p. 159.
6. Ibid., p. 9. Rorty, having made these and similar claims throughout the book, finds it convenient toward the end to disavow the "idealistic" or at least "Kantian" conclusions readers might well have drawn. He does not wish us to think he means that "we make objects by using words" (ibid., p. 276). But, since he has made so much of Sellars's attack on the Myth of the Given, Rorty has no right to imply that *objects* are given. The whole drift of his argument up to that point requires a kind of Kantian view of experience, one at least as Kantian as that of Sellars, rather than the commonsensical one he

The sixth dogma leaves the consensus with a troubling question: granted that no rational account can be given of the nature of the nonpropositional as it might be, so to speak, before the formative power of our rationality intervenes to endow it with propositional reality, precisely what role does the nonpropositional play in the knowledge we do in fact have? We may approach the question by considering the profound distinction between the entertaining by rationality of mere propositions and its entertainment of what, according to the sixth dogma, is the propositional outcome of its own formative activity when, by way of the senses, it deals with the nonpropositional. One instance must suffice: on the one hand, a set of propositions devised by a pomologist to characterize 'Winesap apple' and then entertained by a person who knows apples but does not have a Winesap available; on the other hand, that same person entertaining a Winesap (inspecting it thoroughly and maybe even eating it). In the latter case the person is in a position to enunciate at least a rough equivalent of some of the propositions set forth by the pomologist, although there may in fact be no reason to do so. On the basis of the sixth dogma, the cognitive presence of the apple inspected and eaten—or, as I prefer to say, its presence to rational awareness—is the outcome of a formative rational activity that is propositional (or linguistic), even though what now purports to be present is an apple rather than a set of propositions about apples. Something or other nonpropositional makes the difference, and the difference is readily apparent to the person who inspects and perhaps eats the apple.

What role precisely does the nonpropositional play in thus making the difference? Evidently, according to the consensus, no role that rational awareness can specify, except to say that it does in fact make a crucial difference. And indeed it is not quite clear on what *rational* grounds even that conclusion can be reached. It is this odd situation that causal

---

now falls back on. For Rorty, it must surely be true that objectivity is linguistic in origin. See Chapter 4, Section 5, however, for a discussion of what I call the spurious Kantianism that is now so common within the consensus.

theories of reference try so unpersuasively to cope with, not in the interest of overturning the dogmas we have so far looked at, but in the interest of finding, within their framework, *some* tie to the nonpropositional that shall be rationally expressible—preferably in propositions consistent with physicalism. There is some real thing out there—so say such theories, though we in our present neutrality call it merely the nonpropositional—that causes among other things that peculiar tangy sensation in one's taste buds that is so different, and so welcome a change, from that subtle, blossomlike affection of the same organs that is caused by the ubiquitous Delicious.

The shortcomings of the causal theory of reference, from the standpoint of the dogmatic consensus, are well known: reference may be fixed, but not intension, or *Sinn;* and reference without *Sinn* is blind. To say *what* is indeed there, producing one rather than another effect, we must enter again a complex body of theory, a body of which the theory of taste buds is but a minuscule part. What is really there—that is, nonpropositionally there—is a mystery for the dogmatic consensus except as it is a complex of other *propositional* realities, propositional outcomes of situations in which the formative power of rational awareness meets head-on with the nonpropositional. The very causal relation invoked to fix reference belongs, for members of the consensus, to the same breed: a thing of theory, a propositional reality.

VII. THE DOGMA OF THE INEFFABLE EMPIRICAL STIMULUS. When we attend to what we take to be independently existing concrete particular things, or even a concrete world, rather than terms and propositions, the rationality of our focus terminates in the propositional outcome of our own formative activity. Whatever else enters into our awareness to tell us that we are not dealing with mere propositions enters at a subrational level. The nature of the insistent presence of the nonpropositional in the propositional outcome of the formative, or constructive, power of our (empirically engaged) rationality cannot be rationally expressed.

We are back with the given (sense 1), which was distinguished from the more familiar given (sense 2) in the preliminaries to the sixth dogma—back with the inaccessible raw material of Goodman's worldmaking, the paradigm-independent stimuli of Kuhn,[7] the "non-verbal stimulation" or "surface irritation" with which, according to Quine, ontological issues make connection through "a maze of intervening theory."[8]

The sixth and seventh dogmas lay it down that whatever can be entertained by rational awareness is propositional (or linguistic), but that it lacks any nonpropositional, and thus empirical, footing for the *rationality* of its propositional structure. In other words, the rational structure of our propositional *experience* does not give us the structure that the nonpropositional, *as nonpropositional,* has—if indeed it has any structure on its own. For consensus antirealists, that nonpropositional something-or-other, which functions as a stimulus to cognitive progress, and which is presumably such that our propositional response to it should not be capricious or arbitrary, remains in the strictest sense ineffable—ineffable not only in respect of any structure it might have on its own but also in respect of the nature of the stimulus it provides. If they find it desirable to invoke the honorific old term 'reality', they therefore give it a Pickwickian sense, acknowledging that they are dealing with a *propositional* reality, a *rationally formed* reality, a *linguistic* reality; they thus make a virtue out of giving up any claim to reality—reality without qualification.

But many consensus members are unwilling to settle for this: the true metaphysical nisus, that ineradicable appetite for reality without qualification, is widespread within the consensus. Its target, furthermore, is the obvious one—the one I

---

7. Thomas S. Kuhn, *The Structure of Scientific Revolutions,* 2d ed. (Chicago: University of Chicago Press, 1970), p. 193. See also Kuhn, "Second Thoughts on Paradigms," in Frederick Suppe, ed., *The Structure of Scientific Theories* (Urbana: University of Illinois Press, 1974), pp. 459–82.

8. W. V. O. Quine, *Word and Object* (New York: Wiley, 1960), pp. 276.

am calling the nonpropositional. The tension thus generated within the consensus manifests itself as the realist-antirealist controversy. Consensus realists argue that, although what I am calling rational awareness is incapable of attaining the real, the real is effable enough: a certain ideal body of propositions, those comprised in a perfected body of physical theory, will constitute knowledge of the real. Since they accept the sixth and seventh dogmas as unreservedly as antirealists do, their grounds for making this claim—as distinct from the worth of the claim if quite different grounds could be provided—are not persuasive. The realism-antirealism debate within the consensus is, however, much less tidy than the preceding contrast suggests. There are, for instance, influential antirealists who, although they ought to be willing either to give up reality altogether in the spirit of an unflinching pragmatism or else to settle for a propositional (Pickwickian) reality, nonetheless reveal in some of their occasions a hankering after the same physicalist reality that concerns avowed realists. There is every sign that the realism-antirealism debate cannot be settled if it is left within the consensus.

3 Two Consensus Half-Truths about Science

It must be conceded at once that the dogmatic complex contains at least two truths about science, truths which are, however, so blurred in the setting of the consensus that it is best to call them half-truths. First, much of the work of scientists is concerned with propositions and with propositional structures, that is, structures that include propositional features. Second, propositions and propositional structures are in part the outcome of the formative rational activity of scientists. To remind ourselves how this is so, let us return for a moment to the little catalogue used to illuminate the rubric 'propositions' in the discussion of the first two dogmas.

That catalogue comprises theories, doctrines, propositions, hypotheses, models, formal systems, mathematical objects in general, and (formulations of) laws. In a sense of 'theory' more general than the one that appears in the catalogue, a body of theory may comprise all the catalogue items and more besides. In this sense of 'body of theory', it is quite clear that the activity of science moves in a constant cycle—called in Chapter 2 the theoretic-empirical cycle—between a body of theory and the mode of experience/reality some particularities of which tend to confirm or disconfirm it. And it is also quite clear that this body of theory is at least in part the outcome of the formative activity of the rationality of scientists. As for that mode of experience/reality, it is in effect the world of common sense: it is there that meters and computer screens are read, there that photographic plates are examined and the angles of particle tracks measured, there that a sophisticated staining technique reveals a significant neural connection.

I say that a body of theory is at least in part the product of rationality because I would maintain that any well-constructed body of theory, even one that fails some critical confirmatory test, is constructed out of features or aspects of the real that are independent of our rationality yet nonetheless entertained by our rational awareness. We shall eventually return to that issue, but it lies outside the scope of this chapter. What I want to emphasize just here is that, whatever else goes into a body of theory, the formative or constructive rational activity is also vital to it. So even a discarded body of theory is relevant to the point I want to make.

The first thing that strikes the disinterested observer about the theoretic-empirical cycle is the prima facie distinctness of the body of theory from the experienced world in which it is put to the test. That prima facie distinctness is sufficient to suggest that, although the body of theory may be said to be about that world and may even be thought to display its deeper meaning, the body of theory and the commonsense world are apprehended in quite different ways. We cannot, for instance, know a body of theory at all without having some sort of knowledge of the world that forms the background for the theory—so, at least, it seems before certain philosophical theories are adduced to persuade us otherwise. And it also seems that we know that world, although no doubt less adequately, even if we have no knowledge whatsoever of the body of theory. Knowledge of a body of theory, moreover, gives us indirect knowledge of inaccessible—or at least not immediately obvious—features of the commonsense world the theory purports to elucidate, in contrast with the prima facie directness with which we know that world itself. By "prima facie directness" I mean at this point that knowledge of the commonsense world does not *seem* to be either (a) the entertainment of a heretofore unnoticed body of theory-cum-language, an entertainment that is in fact our way of responding to stimuli that we neither know directly nor experience directly, or (b) the entertainment of cognitive intermediaries (concepts or ideas, for instance) or experiential intermediaries (impressions or sensa, for instance) that permit us to infer or postulate the existence of commonsense entities. When scientists bring a body of theory to an empirical test, they do nothing that sets aside this prima facie status of the commonsense world. This does not mean that *philosophical* theories which question that status need not be addressed; it means only that there is nothing in the practice of science that can be invoked to make such philosophical theories plausible.

But for the moment it is not the contrast between directness and indirectness that most concerns us but rather the contrast between the formative role reason clearly plays in producing a body of theory and the quite different role our rational awareness seems to play in entertaining that part of the commonsense world which bears upon the acceptability of the body of theory.

Suppose that we are listening to a physicist expound the role of the psi-function in quantum theory and that we are competent enough in such matters to follow the mathematics and to take part in the discussion at a professional level. Meanwhile, we are attending, although not in the same way, to such physical experiences/realities as the marks made on the chalkboard, the chalkboard itself, the surrounding room, and the trees outside the window. But our most intense *rational* attention is given to the body of theory propounded by the physicist. Let us call what is thus propounded a complex of propositions, always with the reservation that a body of theory may contain some things—diagrams, for instance—that do not seem to be reducible to propositions. The attention we continue to give the chalk marks and the rest is transformed into a background attention to just the extent we are able to attend competently to the complex of propositions.

As the discussion focuses on the psi-function, we are thus entertaining something that is distinguishable from the chalk marks and also from the particular character of the notation we are using, although—we must suppose—it is not available to be attended to without *some* appropriate notation. The precise ontological status of that something-or-other we are then entertaining, or attending to, is a matter of subtle debate. Indeed, I chose the example just because the wave function, although well understood as a mathematical structure whose predictive power is unquestionable, has not had an agreed-upon ontological status assigned to it. Whether or not it represents something *in natura* more or less adequately, it seems likely that, when we attend to it in the way postulated in my example, we are attending to something that is conceptual as well as propositional. On the other hand, many philosophers now believe that concepts do not exist. We do not have to settle that question—still less the question whether there is some extratheoretic, extrapropositional, extraconceptual isomorph of the physicist's psi-function—to be tolerably sure that our entertainment of the psi-function differs in some fundamental way from our entertainment of the chalkboard. As we entertain the psi-function our attention terminates in something that is sustained by the knower in a radically different sense from the sense, if any, in which the chalkboard is sustained by the knower. If my attention to the chalkboard lapses as I labor to follow what is being written on it, its existence obviously does not lapse for me. And certainly if the word 'chalkboard' never comes into the conversation, the chalkboard itself is no less available to the company. If, on the other hand, I stop attending to the psi-function through some failure in competence or concentration, the psi-function is gone for me, and another competent rational act is necessary if I am to be able to entertain it again.

The psi-function purports to illuminate something in the fine structure of (among other things) the chalkboard. That something, we must suppose, may differ in important respects from its representation with the help of the psi-function. But, whatever it is, it remains as a property of the chalkboard even when the conversation lapses. The expression 'psi-function' in context illuminates the fine structure of the chalkboard by way of something that is entertained by us and is distinct from the fine structure of the chalkboard: when scientists revise the theory they do not revise the fine structure. This point is independent of the question how instrumental intervention to measure the predictions of theory at the quantum level affects what is measured. And it is also independent of the ultimate outcome of the question whether what is measured has what theorists now call local reality, with the preservation of Einsteinian separability, or whether a nonlocal reality prevails.[9] The answers to questions

9. It is sometimes thought that if nonlocal reality indeed prevails, the principle of Einsteinian separability will have to be abandoned, but I do not think that is so. That principle excludes the propagation of a certain kind of causal influence at a velocity greater than that of light. Nonlocal reality simply demands that we understand how two nonlocal realities affect each other without violating Einsteinian separability—that is, how they might affect each other in some way other than by signal propagation. It might turn out, for all we know, that they do not affect each other at all; instead, both might be subject to a correlating influence that is not propagated. For a claim about a parallel situation at the macroscopic level, see my discussion, in Edward Pols, *Meditation on a Prisoner: Towards Understanding Action and Mind* (Carbondale: Southern Illinois University Press, 1975), of the relation between the ontic power of an act and well-separated neural events in the infrastructure that is asymmetrically identical with the act.

of that kind do indeed modify the total structure of quantum theory, and they thus modify the way we *understand* what that theory is about. But the *theory* does not modify the fine structure of which it provides some understanding.

Nor does the fact that rationality plays a formative role in the production and evolution of theory give us the slightest reason to suppose that the experience/reality against which we test theory is a product of that same (allegedly linguistic) formative capacity. Earlier in this century, in the neo-Humean setting of positivism, the power of reason's theory-forming creativity to cope with an experience thought to be radically separate from it remained a puzzle for those dominated by that setting. If the only alternative to that setting were a Kantian one, there might indeed be some reason to take the common world to be a product of linguistic worldmaking. But it is far from being the only alternative; in any event, I hope to show in the next chapter that the "Kantianism" of some factions of the linguistic consensus is spurious.

4 *Intimations of the Directness of Rational Awareness*

It is a major theme of radical realism that what has so far been called prima facie direct knowing of the commonsense world is authentically what it purports to be; but we shall take up that theme only after we have left this *via negativa*. Just here, the consequences of the theme for science may be intimated by an expository suggestion that is still largely negative in spirit. It is this: the true nature of our prima facie direct knowing of the commonsense world is not in accordance with what the sixth dogma lays down. That direct knowing is a mode of rational awareness from whose distinctive character we are distracted by such terms as 'formative', 'constructive', 'productive', 'constitutive', 'meaning giving', and 'worldmaking'. And this is so even though rationality does indeed have a vitally important formative function of the kind seen in theory building, technology, and the arts. With this negative suggestion before us, we may return to the earlier suggestion that a body of theory gives us indirect knowledge of nature, in contrast with a more direct knowledge of nature, and that it is by virtue of prima facie direct knowing that theory is put to the test.

Clearly we know the fine structure of the chalkboard indirectly and by way of the body of theory called into being in and with our use of the language of theoretic physics. It is also clear that, at least at the level for which the psi-function becomes relevant, the fine structure of the chalkboard is in principle beyond the scope of our direct knowing. The chalkboard itself, however, is of the same scale as ourselves and our action, and though from the perspective of the dogmatic consensus our apprehension of it may seem to fall short of the standards for knowledge, it is surely direct enough in a prima facie sense. What is more surprising, propositions can also be known *as propositions* with the same prima facie directness. And any considerable complex of propositions—say a body of theory—can also be known *as a body of theory* in the same way. No doubt propositions in general can be thus directly known only because the formative activity of rationality has first brought them into being. But, once produced and then sustained in being by rationality, they seem to be *there*, wherever such things are, and available to stand in the focus of our rational awareness.

Some in the consensus will find our knowledge of the chalkboard merely phenomenological or macroscopic—one example of the manifest image. And some of them might go on to say that that is what is meant by calling the chalkboard a propositional reality of a certain kind—just as the fine structure, as known by way of physics, is another kind of propositional reality. But even within the consensus, it is now seldom claimed that an expression like 'chalkboard', when used in the presence of a chalkboard, calls into being an intermediary of some sort—concept, idea, representation, or some other *ens rationis*—and that this intermediary rather than the chalkboard itself is then the object of our prima facie cognitive attention. The consensus, which was so long bemused by the Cartesian form of the representative theory that its members came to suppose everybody believed it, has lately come to see that the theory will not do. Though I agree that it will not do as the basis for an account of knowledge in general, especially not for what I am calling direct knowing, that does not mean that a body of theory may not more or less adequately represent what it gives us indirect knowledge of.

The role of sensation in our direct knowing of propositions is, to be sure, different from its role in the kind of prima facie direct knowing that is relevant to the testing of a body of theory. Just here it is the latter kind that most concerns us. We seem to be safe enough in concluding this much about it: although propositions in general (and thus bodies of theory) are products of the formative activity of rationality, there is no plausible reason to suppose that this is true of our commonsense world (experience/reality), which we focus on with what seems to be a direct rational awareness, and which we use to test a body of theory. The point can also be made in terms of language. If a body of theory is called into being with the help of our natural language, and if that theory can then legitimately be said to form part of an extended linguistic net, the function of the body of theory is nevertheless different from that of the natural language in which it is embedded. For we create theories and then make them the objects of our (direct) rational awareness because we cannot attend thus directly to what the theory is about, but natural language is continuous with and completes our direct rational awareness of what it is about.

We are therefore in a position to set aside one contemporary epistemological overstatement about a linguistic net that includes science—the claim that the general mode of experience/reality some particularities of which tend to confirm or disconfirm a body of theory cannot be had independently of the theory. This is in effect the claim that the (commonsense) things some features of which a body of theory purports to explain or elucidate are themselves a function of the theory in the sense that they are functions of the language of which the theory is a part. On one interpretation, this is tantamount to the extravagant and implausible claim that commonsense experience/reality is a function of a body of scientific theory. But, on another and more plausible interpretation, it collapses into the unexceptionable but trivial claim that our confidence in commonsense experience/reality is eroded by our acceptance of a theoretic reality thought of as lying behind it.

Despite wide acceptance of the less plausible interpretation, the experience/reality relevant to a very general body of theory is unaffected by the theory in question. A scientist who understands astrophysical theory about black holes and who is predisposed to accept it—in part because of its fit with existing physical theory of wider import—has exactly the same capacity to read the numbers on some computer screen that tend to confirm or disconfirm the theory in question as does some reasonably intelligent person who is quite ignorant of the theory and quite neutral about its fate. The experience/reality that tends to confirm or disconfirm a body of theory is not theory-laden; it is the import of the experience/reality from the perspective of the body of theory that is theory-laden. That a scientist with a professional stake in a certain theory may dismiss an inconvenient reading as irrelevant or mistaken is an interesting fact about human psychology, but it does not tend to establish a general epistemological doctrine.

The question whether prima facie direct knowing is what it purports to be is identical with the question whether we do in fact enjoy a rational awareness of a reality that is independent of our formative powers. Although we must suspend further development of this question until Chapter 5, it should already be clear that we shall not be relying on arguments that are independent of what purports to be (direct) rational awareness: we have no other resource than the further development of the double epiphany discussed in Chapter 2. But it is wise to set aside a possible misunderstanding of what is at issue before returning to the *via negativa* of our examination of the linguistic consensus.

Recall that prima facie direct knowing comprises our way of knowing the chalkboard, the discoursing physicist, the environing room, the trees outside the window, and the environing world. Because I have in effect been making the point that knowledge does not inevitably focus on or terminate in propositions, it might be supposed that I am trying to call attention to a nonrational or prerational grasp of the world of the chalkboard and the discoursing physicist, a grasp more primordial than propositional expression. That is not what I mean. Direct knowledge may or may not be expressed in propositions, but it can always give rise to propositions when the occasion calls for propositional expression. Thus, if questioned about the discoursing physicist, one might well respond, "That is Wigner, and he is writing something on the chalkboard as he talks, and what he is saying and writing concerns the psi-function"; but our direct knowledge of what is going on was no less rational before being expressed in that way. Moreover, the scope of our direct knowing includes much besides the class of more or less commonsense items mentioned above; most important for our present purpose, it includes propositions and complexes of propositions—not to speak, just now, of all sorts of more imaginative constructs that often occur in conjunction with propositions. So, far from being either antipropositional or prepropositional (in some primitive sense), direct knowledge is precisely what we exercise every time we entertain a proposition as such—what we must exercise, for instance, if someone should ask us to consider the (false) proposition 'Wigner teaches physics at Bowdoin'.[10] The contrast between directness and indirectness

10. I have used the notion of directness in earlier publications, but certain other distinctions I was also concerned to make may have obscured the point I make here. Thus, in *The Acts of Our Being* I distinguish between primary and secondary cognitive engagement. Our rational awareness of the discoursing physicist is an example of primary cognitive engagement, whether or not that awareness is given rational expression; our rational

is not a contrast between what obviously affects our senses (as a chalkboard does) and what does not (as the proposition just mentioned does not, although presumably we are never aware of a proposition as such without some involvement of our senses). It is a contrast between knowing something by virtue of attending to *it* and knowing something by virtue of attending to *something else*. Thus, a competent physicist may be said to know the psi-function directly and, by virtue of this direct knowledge, its theoretic context, and some relevant observation or experiment made in the commonsense world, to know something about the fine structure of the chalkboard indirectly.

Direct knowing is not adequately characterized by the instances I have given of it; it also includes my knowledge that I am now writing and the reader's knowledge of reading in a different "now." Directness is in no prima facie conflict with reflexiveness. As for the relation between direct knowing and common sense, although many of the instances given are commonsense ones, others are not; so direct knowing is by no means identical with commonsense knowledge. Indeed, commonsense knowledge is full of instances of simple and prescientific uses of the theoretic-empirical cycle and therefore rich in instances of indirect knowledge.

---

awareness of the psi-function or of the (false) proposition 'Wigner teaches physics at Bowdoin' is an example of secondary cognitive engagement. But secondary cognitive engagement of this kind is an instance of direct knowing no less than primary cognitive engagement is; both are to be distinguished from the indirect knowledge we have of the fine structure of the chalkboard. I used the notion of direct knowing as early as *The Recognition of Reason* (Carbondale: Southern Illinois University Press, 1963), but there too the point I have just made does not emerge clearly.

4

- Realism versus Antirealism: The Venue of the Linguistic Consensus

1   *The Epistemic Triad and a Dominant Question about It*

- The philosophic position of which this chapter is a partial expression shows science to be realistic in certain circumstances and antirealistic in others. But the position is not so irenic as this beginning might suggest. For one thing, the radical realism on which this view of science depends is different from any of the versions proposed over the years by the forces of realism in that controversy. For another, the antirealism envisioned for some of the multiform occasions of science is not in the least a version of the well-known antirealism of the linguistic consensus but rather a modified antirealism that paradoxically depends for its clarity and force on that same radical realism. The present chapter makes the point that it is impossible to reach a just settlement of the realism-antirealism case—whether construed narrowly, as an issue for philosophy of science, or broadly, as an issue for the entire rational enterprise—if it is tried in the venue of the linguistic consensus.

   Analytic-linguistic philosophy of science has been much concerned in the course of this century with a triad consisting of (a) an accepted body of scientific theory, (b) reality, and

(c) experience, and with the various relations that may or may not exist between the members of this epistemic triad. An accepted body of scientific theory, let us assume, is something comprehensive enough to approximate scientific orthodoxy at a given time. It therefore comprises assumptions, principles, hypotheses, theories (in a sense narrower than what I intend by 'body of theory'), formal systems, equations and mathematical objects of other kinds, models (both mathematical and physical), and no doubt many other things.

During the same period, analytic-linguistic philosophy has considerably widened the notion of language we derive from our acquaintance with natural language. It has done this by conflating that notion first with the notion of a formal system and then with several of the notions in terms of which I have defined 'accepted body of scientific theory'. The wider sense of the term 'language' generated by this gradual but relentless conflation is the one invoked when it is said that science is a kind of language net, or that it is part of a language net more extensive than science. Let us therefore call the first member of our triad an accepted body of scientific language-cum-theory. This allows us both to acknowledge the terminological preference of the linguistic consensus and to signal our unwillingness to accept that preference as holy writ. No doubt the repetition of this awkward hyphenated expression will become tedious; I trust the gain in clarity will justify it.

It was inevitable that discussion within the consensus of the relations between the members of the triad should be conducted from the viewpoint of language-cum-theory—more precisely, that our epistemic condition should be understood to be one of linguistic-cum-theoretic enclosure. A less tendentious way of making that point is this: the consensus assumes that any rigorous expression of the relation between language-cum-theory on the one hand and either experience or reality on the other depends upon the provision of an adequate semantics for whatever body of language-cum-theory is under discussion. Argument about such matters within the consensus has been brisk: consensus realists prefer a semantics—usually some version of Alfred Tarski's—in which the notions of reference and truth are central and the notion of correspondence is not entirely repudiated; consensus antirealists, or nonrealists, prefer a semantics—Michael Dummett's, for instance—in which reference is either repudiated or reinterpreted as internal to the language and the notion of correct assertibility displaces the notion of truth. Whichever semantics is favored, the focus is not on the having of experience on the part of the presumptive knower—still less on any supposed encounter between the knower and the real—but rather on a certain relation that may or may not exist between certain propositions (or statements)[1] of the language-cum-theory and whatever determines the truth value or degree of assertibility of those propositions. And the way to what determines truth or assertibility (whether we call it experience or reality or both) lies *through* language: that, at least, is accessible to the knower; the rest is problematic.

The dominant question about the relations between the members of the triad is the one the realism-antirealism debate within the linguistic consensus has made familiar. The formulation of the question will be made easier if we keep in mind that there are some important matters about which the disputing parties agree, matters not usually made so explicit as I make them, probably because they are consequences of dogmas deeply imbedded in the consensus. First, there is a minimal realism that pervades the consensus. Even those antirealists who prefer not to use the term 'real' acknowledge that it is the concern of science to deal with what we may make a first approach to under the neutral rubric of the nonlinguistic-nontheoretic. It is that which is distinct from and independent of any accepted body of scientific language-cum-theory but toward which that language-cum-theory is

---

1. Usage varies within the linguistic consensus. I use 'proposition' in this book except where such standard expressions as 'observation statement' are called for.

oriented as toward something that must be coped with, dealt with, or otherwise taken seriously. Without this principle of minimal realism science cannot be empirical. To reject it is to embrace an extreme version of idealism; to accept it is to commit oneself to nothing more extraordinary than the view that science must respect experience, and respect it because there is something in it that has not been formed, shaped, made, or otherwise produced by the linguistic-cum-theoretic power of our rationality, even if that something is difficult or indeed impossible to isolate and express.

Second, there is a minimal antirealism that consensus realists, no less than consensus antirealists, accept. We may sum up one feature of it in this way: important though experience is to science, reality cannot be experienced as such, and so when we undergo or enjoy experience what we undergo or enjoy is not that reality in all its independence. The having of experience is our way of responding to what our languages have not made, but that reality does not come through *in* the experience. There is a good deal of ourselves in our experience; and the reason for this lies somewhere in the relation between a body of language-cum-theory and experience.

Third, it is agreed that, if an independent reality could indeed be known, it would be known only by way of some body of language-cum-theory. It is therefore also agreed that the realism-antirealism debate concerns a relation that may or may not exist between a body of language-cum-theory and reality, and not in the least a direct relation between the knower and the real.

We may now build these agreed points into our formulation of the dominant question itself. It is the question whether any accepted body of language-cum-theory can give us knowledge of a reality that is (a) something distinct from and independent of the body of language-cum-theory; (b) something distinct from and independent of our experience; (c) something on which our experience in part depends; (d) something we cannot experience as such; and (e) something which, if we can know it at all, we can know only by way of some body of language-cum-theory. If this is a fair statement of the matter, it should not surprise us if none of the consensus answers does justice to the complexity and concreteness of experience and to its intimate involvement with all our cognitive acts.

2 A Question Once Asked within the Consensus: What Is Experience Really?

Among the various questions in which our dominant one ramifies is one that is now out of fashion, although it troubled the linguistic consensus in one of its earlier phases. It is the question what experience really is. Granted that there is *something* in experience that is quite independent of whatever creative, productive, formative power we may have, what is that something? What is there in experience as we have it that has the epistemic status of something merely undergone, merely received, and thus not of our own making? If we could locate that factor and characterize it, we should then have isolated that feature of it that makes it so worthy of epistemic respect. What purports to be experience is familiar enough; let us call it prima facie or commonsense experience. The domain of common sense includes much besides experience: common sense has, for instance, constant recourse to hypotheses and theories of an informal and prescientific kind. As for commonsense experience, it has both a prima facie texture—sensory, particular, evanescent, often deceptive—and a prima facie content—particular macroscopic things in constantly changing causal interaction, the knower and other persons (or at least one important aspect of them) being included among those macroscopic things. As to that content, it purports to be also cognitive: for one who enjoys it, prima facie experience is not devoid of its own cognitive significance just because it is also the de facto empirical pole of scientific knowledge. Indeed, prima facie experience must have been originally appealed to because it seemed to tell us what was really the case.

The self-criticism of empiricism goes right back to the ancients, but the more developed form of the question what experience really is belongs to the phase of self-criticism that is one by-product of Cartesian rationalism. The attempt begins with the assumption that real experience can be isolated only if the contribution of our senses (including that of our "internal" sensing) can be isolated in a pure form. It always ends in a familiar impasse. Because some version of Cartesian certainty-directed subjectivism dominates the investigation, and because empiricists cannot consistently invoke the rationalistic fail-safe devices Descartes himself fell back on, the merely sensory turns out to be as much an instance of Cartesian ideas as any more patently mental or conceptual findings would be.

Whatever version of "real" experience empirical philosophers settled on was thus problematic in itself, but it nonetheless undermined the authority of prima facie experience. Hume, for instance, maintained in effect that experience consisted of atomic impressions, and this made it inevitable that prima facie experience of such things as enduring physical objects in causal interaction had to be regarded as something to be accounted for rather than something known to be the case. It was not experience after all but rather a product of organizing responses on our part to what experience really was. For unphilosophical persons, and for Hume seated before his dinner or conversing with a friend, these philosophical considerations did not in the least undermine the practical authority of prima facie experience.

In this century, several versions of supposedly real experience have been developed. Moritz Schlick, for instance, seems to have supposed that there is something about experience as it really is which requires that the language in which we record our having of it should approximate exclamations such as "Here now blue" rather than propositions (or statements) in some more usual sense. For him, and for many other positivists in those days, the commonsense objects of our prima facie experience were logical constructions out of the real elements of experience. But since mid-century, for reasons that are well known, the dominant view about experience within the linguistic consensus militates against our trying to determine what experience really is. Today's linguistic consensus holds, in effect, that what we usually call experience is so informed by, constituted by, made by, laden with, or otherwise qualified by a body of language-cum-theory that it must be regarded as a propositional thing. That view, now often expressed in the formula that experience is not *given*, tells philosophers of the linguistic consensus that real experience, something merely experienced and having no taint of our own formative powers in it, is really not to be had. Rationality (in the guise of the linguistic-cum-theoretic) plays so important a part in shaping our experience that an unambiguous appeal to experience itself cannot be made. Not only is it impossible to find an extralinguistic *archē* in the real, it is also impossible to find such an *archē* in experience.[2]

3 The Consensus View of the Relations between the Members of the Epistemic Triad

The reasons that lie behind this complex dismissal of the question what experience really is determine the main lines

---

2. Some landmarks of this complex development should be cited. Moritz Schlick's "Über das Fundament der Erkenntnis," Erkenntnis 4 (1934), had been a response (in the long run unpersuasive) to Otto Neurath's coherentist attack, in "Protokollsätze," Erkenntnis 3 (1932–33), on the notion that protocol statements (or sentences) could furnish a secure empirical foundation for science. Schlick proposed such odd statements as 'Here now blue', which he called 'confirmations' or 'observation statements', to do the empirical job protocol statements could evidently not do. English translations of both papers appear in A. J. Ayer, ed., *Logical Positivism* (New York: Free Press, 1959), pp. 199–227. The nautical image Neurath used in his article to make his coherentism vivid is well known; by 1960, when W. V. O. Quine took that image as an epigraph for *Word and Object*, the propositional character of experience had become part of the orthodoxy of the linguistic consensus. Sellars's attack on the Myth of the Given provides this view of the nature of experience with a memorable name and reinforces the canonical status it had achieved for the consensus; Wilfrid Sellars, *Science, Perception and Reality* (New York: Humanities Press, 1963), pp. 127ff., 196.

of the answers to the question of the relation between a body of language-cum-theory and experience that are open to the consensus. At first glance these main lines seem paradoxical. On the one hand, the consensus takes it for granted that there is something that is nontheoretic, nonlinguistic, and concretely related to us by way of our senses; that the outcome of this relation is what we call experience; and that both the formation and the testing of any body of language-cum-theory thus depends on experience and therefore, in some degree, on that something. On the other hand, it is also taken for granted that experience cannot be had, undergone, or encountered in a way that is merely concrete, sensory, and nonlinguistic-nontheoretic.

The second part of the paradox expresses the minimal antirealism mentioned earlier: it tells us that experience is not given, that it is not independent of the formative power of language-cum-theory. The first part expresses the minimal realism of the consensus: it tells us with equal assurance that something is given, something that *is* independent of that formative power. The paradox is, however, only apparent, for it is not experience but rather the third member of our triad, reality, that the consensus supposes to be given—although given in an odd sense, because the consensus does not suppose that reality can be *taken as given*. It supposes, rather, that reality is given to be responded to by the formative power of language-cum-theory, and that the only knowing or experiencing possible to us is the outcome of this formative power. So reality, though indeed given, is not given to be experienced as given, still less given to be known as given.

It turns out, then, that an answer to the question of the relation between a body of language-cum-theory and experience that is acceptable to all factions of the consensus determines also an agreed position on the relations between all three members of our epistemic triad. In the view of the consensus, experience arises out of the head-on meeting of nonlinguistic-nontheoretic reality[3] on the one hand, and a body of language-cum-theory on the other, but it does not owe its form to whatever structure or nature the real might have independently of our linguistic intervention but rather to the formative power of language-cum-theory.

4 Consensus Realists as Internal-to-the-Language Realists

Consensus realists concede that no *actual* body of language-cum-theory, taken as a whole, gives us knowledge of translinguistic reality. What makes them nonetheless realists is that they claim either (a) that an ideal body of language-cum-theory that would in fact do that job can in principle be produced, or (b) that the actual succession of the bodies of language-cum-theory produced by scientists approaches knowledge of the translinguistic as a limit. The expression 'internal realism' is usually associated with claim (b), but those who make claim (a) also find their justification in an alleged feature of an ideal physicalist language that, if real, would be, as I shall be arguing later, *internal* to it. I therefore take all consensus realists to be internal-to-the-language realists.

The version of (a) developed by Antipodean realists and their allies in the northern hemisphere is economical in the sense that its underlying materialism (or physicalism) is thought by its proponents to need no defense. They assume that at least one body of language-cum-theory, physics, is on its way to completion; and that even now it already meets the most important reality criterion of all: as a body of language-cum-theory, it excludes any reliance on the notion of the scientist as a subject who experiences and knows, and who indeed propounds that very body of language. For the realist,

3. Dummett seems to suggest that this is merely potential reality and that it depends on the intervention of language-cum-theory for its full actuality; Michael Dummett, *Truth and Other Enigmas* (Cambridge: Harvard University Press, 1978), pp. 18–19, 229–30.

that exclusion both explains the empirical success of physics so far and is a precondition of its ultimate completion. The obverse of this metaphysical assurance on the part of consensus realists about what reality *is* is their conviction of the inherent spuriousness of prima facie experience considered as a report on the real having cognitive significance, and of the spuriousness of the notion of an experiencing and knowing subject that goes with it. Their monolithic agreement on these issues makes their disagreements (for instance, about whether materialism is properly reductive, eliminative, or functionalist) much less momentous than they appear. We may therefore discuss the movement in terms of a generalized version of it. For all such realists, one member of the epistemic triad, experience, together with its correlate, the subject who undergoes or enjoys the experience, is the villain of our epistemic condition; and it becomes the duty of the philosopher (as distinct from the scientist) to show how, in principle, the villain can be exorcised from a body of language-cum-theory. To put the matter another way, these realists suppose that, although at least one body of language-cum-theory—that of common sense—does in fact give rise to a correlative mode of experience that purports to have cognitive content, it is not the proper mission of a body of language to produce so-called experiential knowledge but rather to provide pure knowledge about what cannot in principle be experienced as it translinguistically, transtheoretically is.

We must in due course consider the consensus realist claim that commonsense language gives rise to a correlative mode of experience. Just here it is important to notice that realists do not generally make this claim in a Kantian spirit. There is, however, an important exception, Wilfrid Sellars, who propounded in the 1960s a view of commonsense experience—and of philosophy which takes commonsense experience seriously—that is much in debt to Kant for its spirit, if not for its details. What purports to be a cognitive and experiential awareness of ourselves as persons is in fact our entertainment of a "manifest image," that is, a world of appearance whose articulation depends on our linguistic response to reality. For Kant, there is no *cognitive* circumvention of whatever formative powers we may deploy in constituting our commonsense world. Sellars, however, argues that the propositions of science give us reality in a sense in which the appearances of the manifest image do not: "it is 'scientific objects', rather than metaphysical unknowables, which are the true things-in-themselves."[4] It is, of course, a science of the future Sellars has in mind, a science that can in principle come to pass (even though it may not in fact do so) and therefore can give us now at least an ideal for what a cognitive grasp of reality would consist in. Sellars argues that an ideal picturing (representing) relation exists between the ideal scientific language and the nonlinguistic items it illuminates; and he seeks to naturalize that picturing relation, so that it can in principle be studied like any other thing science studies. It is, he thinks, a matter-of-fact relation, whose genesis can be studied "in the order of causes and effects, i.e., *in rerum natura*."[5] That extralinguistic items (or linguistic items functioning, qua objects of study, as extralinguistic items) cannot be rationally encountered in experience has been laid down by the sixth dogma. Since Sellars has contributed so much to that dogma, by way of his attack on the Myth of the Given, one hopes that he himself was clear on that point, and that it is rather the relation of an ideal physicalist language of the future to our other languages—in particular its in-principle supersession of commonsense language—that endows it, in his view, with the capacity to pronounce upon the thing in itself.

Many recent consensus realists who are not Kantians even in the incomplete sense in which Sellars is a Kantian—Hartry Field, for instance—and who might quarrel with Sellars's use of the notion of representation, nonetheless try, as Sellars does, to naturalize the relation between a body of

4. Wilfrid Sellars, *Science and Metaphysics* (New York: Humanities Press, 1968), p. 143; see also p. 50.
5. Ibid., p. 137.

language-cum-theory and reality by way of a causal theory of reference. Only a physicalist causal chain, they think, can put a body of language in touch with nonpropositional reality and so ground the truth of that body of language. These "semantic" physicalists thus share with earlier consensus realists the notions that realism becomes the more plausible as conformity with the fifth ("no subject") dogma becomes more rigorous and that the correct description of the nonpropositional in itself must be a physicalist one. In all consensus realists there is a disposition to suppose that some one of our languages—one we are on our way to developing—will be so profoundly right that it will give us more than a merely propositional, a merely Pickwickian, reality.

Still deferring discussion of the dominant consensus realist view about how—as they allege—commonsense language gives rise to experience, let us note that it is already clear that the most important realist criterion for an ideally sound body of language-cum-theory is a negative one: that ideal language must not generate, for one who holds or uses it, the delusion of being a subject veridically experiencing a world. It must be capable of explaining whatever happens in nature, including the (apparent) conscious intelligent action of human beings, their (apparent) power of introspection, and their capacity for language-cum-theory. But its internal propositional structure must be utterly free from any reference to, or other dependence on, the notion of an experiencing subject. Because so-called intentional language is thought to imply the presence of a conscious (or at least meaning-giving) subject, this kind of consensus realism is unambiguously clear about one point: whatever structure the perfected language that is to give us knowledge of reality is to have, it must have no propositions characterized by intentionality.[6]

6. The minimum consensus view about intentionality is that it is a property of certain complex propositions which express a propositional attitude—belief, for instance—toward some component proposition. It is minimum in the sense that some consensus members suppose it can be held without commitment to such notions as mind and consciousness; see, for instance, the discussion of Putnam's internal realism later in this section.

Indeed, it must not have persons, knowers, and conscious beings as part of its ontology; it must not postulate entities of that kind.[7] What has heretofore been regarded as the knower, as the one who undergoes or enjoys experience, as the subject, is to be absent as such. In that future body of language-cum-theory the subject will appear as what it really is: a complex physical system in causal concourse with other physical systems under whatever laws ultimately turn out to prevail in such systems—the *real* laws of nature. The inherent spuriousness of prima facie experience would then have become intelligible in terms of purely physical laws.

The contrast of a bad (or delusive) body of language-cum-theory with a canonical body of language that expresses (or explains) the translinguistically real is the heart of the doctrine of linguistic-consensus realism. But how can we demonstrate any such thing while continuing to adhere to the general principles that pervade the linguistic consensus? It might well be that realism with respect to science is compatible with other principles—those of radical realism, for instance—but that is not the question now before us. Turning our attention to that, let us consider the alleged relation between the commonsense body of language-cum-theory and experience. Consensus realists claim that commonsense experience (which is in some sense sensory or perceptual) is so laden with that body of language as to be spurious as a reality report. This is by no means just the claim that a scientist committed to a certain theory may fail to report some particular instance of commonsense experience as it would have been reported by someone not committed to the theory. That happens often enough, especially when the perceptual situation is ambiguous. Thus Percival Lowell, seeking to confirm his theory about the surface of Mars, sees canals, and sees them so well that he is able to draw an elaborate map of them.

---

7. The parallel in today's linguistic consensus to the positivist doctrine that commonsense objects are logical constructions out of the real elements of experience is the doctrine that commonsense objects belong to the ontology of commonsense language.

Others, neutral about or hostile to that theory, see the surface differently through the same telescope. Evidently something conceptual is interfering with that instance of perceptual experience in Lowell's case, and we may concede at once that it is the particular theory he holds about the surface of Mars. But the claim consensus realists are making is much more radical: they claim that commonsense experience, considered as a general mode of experience, is just as theory-laden as was Lowell's particular instance of commonsense experience. Yet they offer no argument for this radical claim except (a) other examples analogous to the Lowell case and (b) a reductive argument for the elimination of the subject that assumes in advance the spuriousness of any subject's commonsense experience.

Except for Sellars, they do not rest their case on a Kantian argument for the phenomenal character of ordinary experience, and so we can only conclude that most of them are saying that what we ordinarily call perception of the commonsense kind is in fact a kind of *conception:* we misinterpret, by way of concepts that are integral with commonsense language, what are in fact physical happenings within our environment, within our central nervous systems, and between our environment and our central nervous systems. This theoretic misinterpretation, this muddled conception, is not alleged to be the *cause* of a corresponding muddled perception; it is instead alleged to be *identical* with ordinary perception. So the doctrine does not really claim that ordinary perceptual experience is loaded with theory; it claims that what the speaker of commonsense language supposes to be the perceiving of a world is in fact the holding of a bad hypothesis or theory—often called a folk theory—about a reality that in point of fact is not perceivable at all.

Turning to the question of the inner structure of any body of language-cum-theory, we find that the alleged status of commonsense experience creates difficulties for the notion of reference, which is so central to consensus realism. The difficulty about commonsense reference is obvious enough: it is spurious if there is something spurious about commonsense things. If those things are merely postulated by the commonsense body of language-cum-theory, if they merely belong to its "ontology," they cannot be things referred to. Thus, although 'person' refers to a person in commonsense language, that person's ontological status depends on the language in which 'person' occurs—depends on it in the very fundamental sense that we experience persons by virtue of our use of a language in whose network the term 'person' has a certain place. The meaning of the terms of commonsense language do not therefore derive from the things apparently referred to but rather from the relations of those terms to other terms in that network. Commonsense reference is thus made an internal-to-the-language notion; indeed, consensus realists have adopted toward commonsense reality the same (coherentist) position consensus antirealists adopt toward the reality question in general.

So much the worse for common sense, consensus realists might say; we are only interested in defending the usual semantic status of reference (providing a connection between theory and what lies outside theory and is in no sense generated by it) for the case of a canonical physicalist language. But this dismissal of common sense has repercussions: that usual semantic status of reference is rooted in common sense. So too the usual realist causal interpretation of reference: its persuasiveness depends upon our taking such things as persons and theories and causes, and thus propositions whose truth is correlative with something causally related to the one who entertains or asserts propositions, in something like their ordinary or commonsense acceptation. And this acceptation is precisely *not* what is preserved in the ideal physicalist language-cum-theory. If the term 'reference' is to appear in that physicalist language at all, it must be transformed into a causal relation between physical items; and 'cause' must itself be transformed (presumably into something like 'the earlier of two states described by some sort of continuous mathematical function'). In that sense, physicalist realists must in the long run rest their case on a sense of 'reference' internal to the ideal language, one that no longer refers (in the ordinary realist sense) to ordinary realist reference.

Besides this logical difficulty, which assumes the completion of the physicalist program, there are other difficulties that are clamant right now. One of them is that progress toward the completion of the physicalist program is possible only if scientists continue to rely on experience, and prima facie experience at that. Although allegedly spurious as a reality report, prima facie experience persists as the pragmatic point of departure for theory building, and it continues to lay science under the obligation of bringing a body of language-cum-theory back to it for testing. All this means that each step toward the ideal physicalist theory must be judged not just by its contribution to the eventual elimination of the experiencing subject but also by its success in coping with that same subject's prima facie experience. A bluff, hearty, and everyday realism has always been needed to get a physicalist program under way; and that attitude is so natural to philosophers who choose this option that they may not even be aware of this paradox.

We turn now to the other version of consensus realism, the one that sees in the succession of scientific theories an internal mark of science's asymptotic approach to a body of language-cum-theory characterized by what amounts to absolute correctness of assertibility. The doctrine in question is the one Hilary Putnam has been defending at least since 1976 under the name "internal realism." It has changed considerably since then, and although Putnam has continued to use the same name, the doctrine in its most recent versions belongs clearly with antirealism, as I argue later. But as late as 1982, Putnam was still defending, in a context confined to the status of science, a version of internal realism that has important affinities with the scientific realism we have just been discussing.[8] Putnam's internal realism of that date offers,

8. Hilary Putnam, "Three Kinds of Scientific Realism," *Philosophical Quarterly* 32 (1982): 195–200.

however, no criterion that would single out some actual body of language-cum-theory of the future as the ideal one. He supposes only that there are marks or features internal to the succession of bodies of language-cum-theory that allow us to determine that science is making progress toward absolute correct assertibility as toward a limit.

In the article in question, it is at first Putnam's differences with doctrines like the one discussed above that are salient. In Part 1 of the paper, "Scientific Realism as Materialism," his target is a physicalism like that of Field, which he discusses without calling it, as I think many critics would, a metaphysical doctrine. Field, he says, "would argue that 'intentional' or semantical properties (for example, reference) can be reduced to physical ones." Putnam does not wish to refute Field's doctrine by arguing for a metaphysics that maintains the transpropositional reality of mind, and with it intentionality; the whole tenor of his article is antimetaphysical. His own defense of intentionality (at that time) is merely that there are correctly assertible propositions that contain such terms as 'reference' and that such terms are not reducible to a physicalist language. He confines intentionality, however, to the domain of what I am here calling propositions: "Truth, reference, justification—these are *emergent*, non-reducible properties of terms and statements in certain contexts." His dualism, he goes on to say, "is not of minds and bodies, but of physical properties and intentional properties."[9] It should be remembered that Putnam had earlier been persuaded by Dummett that a reference-free, non-realist, or "verificationist" semantics is more correct than one

9. Ibid., pp. 195–96. This is the minimum sense of 'intentionality' mentioned in note 6. In his more recent work, Putnam adopts a more generous sense of 'intentionality', one closer to that of Husserl. The term thus becomes roughly equivalent to 'mind' or 'consciousness'. This move, now common in antirealism, requires closer critical attention than it has received. In view of the phenomenological origins of the term 'intentionality', to make it equivalent to 'consciousness' is to imply that consciousness forms, shapes, constitutes, or otherwise gives meaning to its objects. It is by no means clear that this is so, and radical realism provides quite a different view of the matter.

in which reference plays the role consensus realists assign to it. But Putnam did not drop the notion of reference. Instead, he gave it the status of a notion that is internal to the language and called the resulting position 'internal realism';[10] in the article I am considering, it is called 'scientific realism as convergence'.

There are important differences between Putnam's internal realism and a generalized scientific realism we might construct from what is common to the positions of, say, Field, Paul Churchland, C. A. Hooker, Robert Nola, Richard Boyd, and Michael Devitt. The similarities, however, remain striking: we saw, for instance, that the ideal physicalist language also turns reference into a notion that is internal to the language. There is no doubt what Putnam wants to do: he wants to steer a course between antirealism of the neopositivist kind and a realism that makes metaphysical claims. But it is by no means clear that he does so. In the section of the article called "Scientific Realism as Convergence," he distinguishes his position from antirealism of the neopositivist kind by way of a doctrine of the preservation of reference across (suitable) theory change. The neopositivist doctrine that theories are "black boxes" yielding successful predictions is thus repudiated in favor of a realism that holds that theories yield "successive approximations to a correct description of microentities."[11] For my part, I think that this is indeed what at least some of our theories do in the case of some microentities. But the question now is whether Putnam's internal realism gives us good reason for thinking so; and, however that may be, whether that doctrine is not open to some of the objections Putnam brings against what he calls, here and elsewhere, scientific realism as metaphysics.

At first glance his claim about the preservation of reference seems modest and pragmatic. It postulates a regulative

10. Hilary Putnam, *Meaning and the Moral Sciences* (London: Routledge & Kegan Paul, 1978); see especially Part 4, "Realism and Reason," in which the change in Putnam's position brought about by the influence of Dummett is effectively summed up.

11. "Three Kinds," p. 199.

ideal, but one that seems practically instantiated: science does seem to tend toward an *actually* settled description of the electron, for instance, and it is plausible that it will be even more stable as time goes on. The theories thus seem to converge on settled descriptions of referents at least in the sense that they gradually yield such descriptions over time. But one reason the description has settled down is the increase in predictive richness (at the macroscopic level of ordinary experience) as we move through the succession of theories in which one term, say, 'electron', purports to refer to the same microentity. So have we any right—seeing that we are working within the assumptions of internal realism—to dismiss a neopositivist interpretation? A neopositivist might well remind Putnam that the term 'referent' is (for internal realism) internal to the theory, as indeed is 'electron', so that the total theory might still be regarded as a prediction device; the postulation of a referent would then be a heuristic device, useful because the ultimate source of the notion is the world of common sense and because our sensory experience does indeed belong to the world of common sense, but not to be taken as having any more realist significance than a logical construction had in earlier positivism.

There is evidently another regulative ideal at work in internal realism: that of *unqualified* correct assertibility—that is, a body of language-cum-theory, and a correlative set of descriptions of referents, that could not in principle be superseded. We may approach this ideal by noting that almost everything would fall into place for the 1982 Putnam if we already knew certain things: that there is a referent there; that it has a stable nature independent of whatever referring we do and whatever formative activity goes into our theory making; that what is correctly assertible about it in a given body of language is closer to it (more correct) than what was correctly assertible about it in each of several earlier theories; and that there was significant referential carryover from theory to theory through the succession. If we knew all these things, we could plausibly claim that our actual theories approach (converge on) an ideal theory as a limit; that our descriptions as a consequence converge on the referent's real nature; and that this is the reason for the settled thread that in fact runs through the succession of descriptions.

Not everything, however, would fall into place. Not knowing the limit, and being unable to recognize it as the limit even if it were propounded to us, we could not be sure how close our latest version was to it. (On the basis of what internal realism has in common with antirealism, some other language—that of common sense—might introduce a prediction-relevant layer of phenomenal experience that limited progress toward the referent beyond a certain point.) So we could not even say that our latest theory and our latest referential description gave us reality for every practical purpose. And there are these final difficulties: if internal realism were true, how could one know all this? And if one did know it, what help would that be to an internal realist, seeing that what one knew in that case would have important features in common with an Antipodean realism?

Since his tergiversation of 1976, Putnam has seemed bent on transforming himself before retirement into the complete Continental philosopher. The internal realism (with respect to science) we have been discussing exhibits the difficulty of taking some decisive steps in that direction while retaining some of the features of standard scientific realism. Putnam's later work belongs without qualification to antirealism, or nonrealism; it is discussed in the following section.

5 Consensus Antirealists as Pseudo-Kantians

The antirealist option makes not experience but another member of our triad, reality, the villain of our epistemic condition. It is not reality in the sense of the minimal realism mentioned earlier that is the villain, but rather reality in the stronger sense of something understood to be both translinguistic and nonetheless attainable and knowable by virtue of *some* body of language-cum-theory. If we give up that reality as a cognitive ideal, antirealists suppose, everything else falls into place. This renunciation even leaves antirealists with an ontology they take to be respectability itself: ontology of language—indeed, of the best body of language-cum-theory current science has to offer. There will thus *be* whatever entities, processes, and events the currently accepted body of language postulates. If it includes subtheories that are in some respects incommensurable with each other but that are both necessary for an adequate empirical account, then something called by one name will be one kind of entity in one subtheory and a different kind of entity in the other subtheory. Thus, in one subtheory an electron is a particle and in another it is a wave bundle. And even if this oddity marks a permanent impasse—even if there is never to be an empirically adequate theory in which an electron is some one *tertium quid*—it seems fair to say that many things that now *are* in the sense that they are part of the ontology of today's accepted body of language-cum-theory will no longer *be* in some body of language-cum-theory of the future. The notion that any legitimate ontology must be the ontology of a language helps encourage one terminological complication for any adequate characterization of antirealism, for title to the term 'realist' becomes ambiguous if 'real' in fact means only 'real for a language'. Thus any antirealist may just as easily claim to be a realist, and indeed Putnam does just that in his more recent work.

The reason for the restriction of ontology to ontology of a language is the failure, in the view of the consensus, of any mode of experience—prima facie or any other—to be experience of the real, where 'real' is understood in the sense antirealism wishes to eliminate rather than in the Pickwickian sense just noticed. And this failure of experience to be what it purports to be (to the naive mind) goes back to the minimal *anti*realism both factions of the linguistic consensus agree on: experience is informed by, laden with, made by, constituted by, or otherwise qualified by a body of language-cum-theory. Minimal *realism*, however, also continues to be respected by antirealists: some ineffable factor in experience brings us up against what we have in no sense made or formed—precisely the nonlinguistic, nontheoretic; so our bodies of language must eventually be brought to the bar of experience to find out their worth. This means that, for the purposes of science, the best body of language-cum-theory is the one that allows us to predict the particular course of experience, no matter how thoroughly the general texture of that experience is laden with theory.

We have already seen that there is an important de facto reliance on prima facie (commonsense) experience in the case of consensus realists. Is it any different for antirealists? Surely, with respect to science, they are antirealists chiefly because they suppose that an accepted body of language-cum-theory is accepted not because it tells us how it is with nonlinguistic reality but because it allows us to anticipate, in some here and now, what we shall be experiencing here or elsewhere at some different time and because it performs this office better than other available bodies of language-cum-theory. Where it fails—Paul Feyerabend gives us many striking examples of the empirical failure of an accepted body of language-cum-theory—it continues to be accepted because scientists assume that either further theoretic development in the indicated direction or more ingenious instrumentation will indeed increase our power of prediction and control.[12] Efficiency in thus saving the phenomena is the antirealist criterion that in the long run justifies the supersession of one body of theory by another, or the supersession of one subtheory by another, or the supplementation of one subtheory by another that is in some sense incommensurable with it. But, to repeat, will the "phenomena" be anything different, in practice, from prima facie experience? To put the question in another way: will the empirical arena in which tests are made be anything other than what is public in the sense that theory holders and those who are not theory holders, scientists and nonscientists, can all agree on it? Is there working in the antirealist side of the

12. Paul Feyerabend, *Against Method: Outline of an Anarchistic Theory of Knowledge* (London: N L B; Atlantic Highlands, N.J.: Humanities Press, 1975).

linguistic consensus any more profound relativism with respect to science than the not very menacing one just outlined? It has already been suggested that, for consensus realists and antirealists alike, experience is informed by, laden with, constituted by, or otherwise qualified by a body of language-cum-theory. So, to answer the questions just posed, we shall have to determine just how consensus antirealists understand this alleged informing of experience.

Terms like 'formative', 'informing', and 'constitutive', it would seem, are more appropriately used (with respect to experience) about antirealism than about realism, for antirealists profess to have something in common with Kant's claim that our prima facie experience is thoroughly saturated with the formative power of rationality. Kant's own doctrine was a creative response to the crisis in empiricism brought about by the self-criticism of empiricism in the persons of Berkeley and Hume. There was a similar crisis in empiricism earlier in this century: the struggle of the positivists to reconcile their own version of Hume's Relations of Ideas (an analytic body of language-cum-theory whose bony structure consists of an extensional logical "language") with their version of his Matters of Fact (sensory "givens" ideally expressed by synthetic statements whose terms have meaning in accordance with the verification principle) rehearses in a more technical and sophisticated way many features of that eighteenth-century self-criticism. It is thus not really surprising that some of the creative response within the consensus to the failure of positivism was Kantian in spirit and sometimes even consciously emulative of Kant. Among the writers responding in this general way we may include Wittgenstein (at the time of the writing of *Philosophical Investigations*), Sellars, Goodman, Strawson, Rorty, and, more recently, Putnam. (Sellars belongs with the realists, and just now we are concerned only with the antirealists.)

But precisely what sort of loading of experience by language-cum-theory do consensus antirealists envision? Do they accept Kant's circumstantial account just as it is? Do they offer an alternative account or accounts? Let us begin our consideration of these questions by speaking of Kant's own doctrine as a *strict Kantian loading of experience* and remind ourselves of its principal tenets. For Kant himself, when we operate cognitively in the mode of understanding,[13] all our sensibility is category-laden (laden with certain pervasive concepts) and those same categories are sensibility-laden. In Kant's own doctrine this mutual ladenness is worked out in detail in terms of what he calls the schematism of the categories. This bond between understanding and sensibility is central to the Kantian epistemology; indeed Kant's claim about the possibility of synthetic a priori judgments depends upon it. That is why our understanding of nature, or at least our understanding of certain pervasive features of nature, is for Kant also an experiencing of nature. Conversely, "experience itself is a way of knowing."[14] But this is a way of knowing that gives us the objective world of Newtonian science, and what is thus known is an appearance that brings with it no report on the structure of the real—in Kant's terms, the thing in itself—but only (after the intervention of critical philosophy) a report on the organizing power of the understanding and the sensibility in dealing with an otherwise ineffable thing in itself.

It is of great importance for our present purpose that this very general understanding-experience is inflexible; that is why it is appropriate to call it a strict Kantian loading. Its inflexibility is such that no version of the laws of nature can be at odds with it, which means that no particular body of theory could load experience in any way that was in conflict with it. But within this inflexible framework, our particular formulations of, or versions of, the laws of nature are flexible enough. Thus, despite the influence of Newton on Kant, there is nothing in Kant's epistemology which requires that the inverse square law, rather than some other gravitational law, should prevail in nature.

13. Kant distinguishes reason from understanding and makes claims about the former that are not relevant to the concerns of today's antirealists.
14. *Critique of Pure Reason*, Preface to the Second Edition (B, xvii).

This flexibility within an inflexible framework has a counterpart in Kant's distinction between the matter and the form of experience. As to its particularity, experience is— what it shall be; we must wait upon it to see how it turns out. The features of understanding and sensibility (linked by the schematism of the categories) that make, in his view, for synthetic a priori knowledge lay it down that when I look out my window I shall see some kind of (phenomenal) substances in lawful causal interaction, but they do not lay it down that those substances shall obey some particular set of laws rather than another; and they do not lay it down that I see (as I now do) trees and grass rather than, say, a rocky shore and the ocean. Whatever laws I may judge to be operative in nature must accord with what the matter of experience decrees: experience will be as it shall be, subject to the very general synthetic a priori constraints Kant envisions.

It is difficult to see what in Kant would lay it down that an advance in our understanding of the laws of physics would call for a change in the mode of experience against which we test those laws, which is the kind of flexibility antirealists, when they are writing about science, seem to require. The most a roughly Kantian epistemology can claim is that our experience is so informed that its structure will always be that of objectivity; and that objectivity, being subjective in origin, will always be phenomenal. Although phenomena are indeed relative to ourselves, they are cognitively absolute in the sense that no invention of a new language can circumvent them. There is no linguistic relativity where a strict Kantian loading of experience prevails.[15]

It is clear that a strict Kantian loading is consistent with what some antirealists say about science. We have only to

15. It might be thought that for Kant the language of morals is an exception, but it should be remembered that the categorical imperative legislates a universal way of acting within the world of phenomena: our obligation is to act in a certain way despite a pattern of inclinations that belongs to the world of cause and effect and thus to the phenomenal world. The whole point of the Kantian doctrine of morality is that our obligation is utterly at

assume that it is the job of a scientific body of language-cum-theory to enable us to predict and in some measure control the course of experience, where 'experience' implies 'phenomenal but ineluctably so'. As suggested earlier, supersession of one body of theory by another would then take place because of the superior predictive value of the later one. The improved theory would not give us knowledge of transtheoretic reality, nor for that matter would it necessarily bring us any closer to it: the structure of the translinguistic might in principle be different from what our formative power imparts to it. And though our ability to cope with experience might be much improved, the texture of experience itself would remain just as it was.

On the other hand, although a strict Kantian loading is thus consistent enough with some versions of antirealism, I do not find that any influential antirealist has argued systematically for it. And there are many influential antirealists—Goodman and Putnam, for instance—who seem to think that the very texture of experience must alter with changes in theory. In any event, to argue for strict loading would create difficulties for the other "languages" in terms of which antirealists now so commonly discuss human activities of a nonscientific kind.

Let us inquire, then, whether antirealists argue for what we might well call a *flexible Kantian loading of experience*. A flexible loading is what would prevail if there were many—in principle, infinitely many—categorial schemes (or linguistic forms, symbolic forms) each of which was in the same intimate union with sensibility which Kant envisions for his own strict loading. If we suppose that these schemes fall into families in accordance with our various interests—some subserving a scientific, others a moral, artistic, or religious interest—and that within the sphere of a given interest any scheme might be replaced by another, then we have an antirealism of an extremely relativistic kind. The many modes of appear-

---

odds with what our science requires. For Kant, in any event, the language of morals does not *make* obligation.

ance generated by our various languages would signal not our enjoyment of many facets of the nonlinguistic real but rather the formative power of our many languages in responding to the givenness-to-be-formed of the ineffable real. An experiential relativism of this kind is what many antirealists seem to think prevails, and many of them rejoice in that supposed state of affairs, taking it as a sign of our freedom and creativity. On this view, if there is indeed a translinguistic-transtheoretic reality, it does not constrain the variety of our response to it. No "reality" (in the sense of the ineluctably phenomenal) constrains us either, and as to the worlds that are experiential counterparts of our multitudinous schemes, it does not really matter whether we call them realities in some Pickwickian sense, or antirealities, or irrealities. The point is that they *are* as we make them to *be*. If it happens to be science that is our theme, then a new body of language-cum-theory immediately generates the structure of its experiential counterpart: our way of experiencing literally changes with our new linguistic scheme.

Is there any antirealist who actually makes out a case for such a claim either about the practice of science or about other rational human activities? Although Kant's name is invoked often enough by antirealists, I have not been able to find any explicit and well-developed argument for a flexible Kantian loading. Probably no philosopher of science makes out a stronger case for the overwhelming of experience by theory—or at least for an autonomy of theory that produces a failure of the distinction between theoretic and observation language—than Feyerabend does. But although it sometimes seems that Feyerabend is supposing that a flexible Kantian loading of our experience is brought about by holding a particular body of language-cum-theory, most of what he has to say suggests instead that the partisan of a theory hides, disregards, overlooks, or otherwise fudges evidence that would be just as available to the partisan of another theory.[16]

16. Feyerabend, *Against Method*, especially chaps. 5, 7, 14.

And that is quite a different matter. It is consistent with Feyerabend's doctrine to suppose that the general texture of prima facie experience persists through theory change, however radical; and that prima facie experience thus remains the empirical standard for both scientist and nonscientist, even though a truly creative scientific nonmethod often finds it convenient to ignore some empirical difficulty in the short term in the interest of a long-term gain.

What about Goodman? Does not he, at least, assume a flexible Kantian loading of experience, and one that operates not just within science but also within each of our other worlds? He says, after all, in *Ways of Worldmaking*, that he thinks of the book "as belonging in that mainstream of modern philosophy that began when Kant exchanged the structure of the world for the structure of the mind, continued when C. I. Lewis exchanged the structure of the mind for the structure of concepts, and that now proceeds to exchange the structure of concepts for the structure of the several symbol systems of the sciences, philosophy, the arts, perception, and everyday discourse." The movement, he goes on to say, "is from unique truth and a world fixed and found to a diversity of right and even conflicting versions or worlds in the making."[17] At the beginning of the text itself, he speaks of the many themes he has in common with Ernst Cassirer, so a reader might well expect the book to unfold against a Kantian (or at least neo-Kantian) backdrop. But this reader, at least, found in Goodman no persuasive *argument* for a flexible Kantian loading of experience. In Chapter 5 there is indeed a discussion of "how perception makes its facts," but it is based on Goodman's report of controlled experiments deliberately designed to produce something perceptually ambiguous, and it is by no means clear that the perceptual "projections" that at least some people make in such situations allow us to infer that perception is in essence projec-

---

17. Nelson Goodman, *Ways of Worldmaking* (Indianapolis: Hackett, 1978), p. x.

tive.[18] But in any event Goodman does not undertake to show that there is the kind of mutual interpenetration of theory and perception that a flexible Kantian loading of experience would require for each language-cum-theory (or each "version"). In Chapter 7, "On Rightness of Rendering," in which he brings together the various themes of the book, his argument about different worlds (or true versions) culminates in an example that belongs to mathematics rather than to sensuous experience.[19] The facts made by virtue of each correct system of that kind exist in the same linguistic-theoretic space in which they are contrived; they are, that is, not presumptive experiential facts like those of the perceptual experiments discussed in Chapter 5 of *Ways of Worldmaking*.

It might seem to be appropriate at this point to ask whether Putnam's version of antirealism provides any argument for either a strict Kantian or a flexible Kantian loading of experience. But some readers may first have to be persuaded that Putnam, who produced in 1987 *The Many Faces of Realism*, is really arguing in that book for what others call antirealism. Putnam has ancient roots in scientific realism, and I argued earlier that his 1982 version of internal realism (with respect to science) belongs under the rubric of consensus realism. Nonetheless, *The Many Faces of Realism* is a fairly standard contribution to the literature of consensus antirealism. Its thesis is not radically different from that of Goodman's *Ways of Worldmaking*, but Putnam seeks to repossess the traditionally honorific term 'real' by calling what for Goodman would be a world (or an irreality) a *reality*. Putnam is thus using the term 'real' in a Pickwickian sense: throughout his book what is real includes what he calls "a contribution made by language or the mind"; so 'realism' is tantamount to what others mean by 'antirealism'.[20]

18. Ibid., p. 89.
19. Ibid., p. 120.
20. Hilary Putnam, *The Many Faces of Realism*, The Carus Lectures (LaSalle, Ill.: Open Court, 1987), p. 8.

Putnam draws heavily on Kant, or rather—surely one may say it to a conceptual relativist without offense—a certain version, or reading, of Kant. Putnam's references to Kant are somewhat offhand, but it seems fairly clear that he supposes the terms 'thing in itself' and 'noumenon' to be interchangeable for Kant. The arguments he then brings against the "intelligibility of thoughts about noumena," arguments which, as he notes, have their parallels in the *Critique of Pure Reason,* he therefore takes to be arguments against the notion of the thing in itself. This leads Putnam to suppose that the notion of the thing in itself is not essential to the doctrine of the first critique; and this train of thought seduces him into the further supposition that his own doctrine is free of any reliance on the doctrine of the thing in itself.[21] Both suppositions are false, and it is important to see why.

For Kant, the restriction of knowledge to the phenomenal, which for him is also the objective, has as its absolute correlative the doctrine of the thing in itself, for he thinks we give the form which is objectivity to something we have had no hand in making, and which, whatever little else can be said about it, certainly does not have *that* form without the intervention of ourselves as knowing subjects. When he says that the thing in itself is any empty notion, he means only that it cannot be an object of knowledge and so cannot be characterized in any positive way. Still, Kant was well aware that we are tempted to make knowledge claims about the thing in itself anyway. When we succumb to the temptation, he thinks, we do so in one of two ways: (a) we try to characterize the thing in itself as we would a (phenomenal) object, treating it as though it possessed in itself what the union of understanding and sensibility provides in making it phenomenal; or (b) we try to characterize it as a noumenon (or constellation of noumena)—that is, we suppose that it is a being (or beings) of reason and suppose ourselves capable of deploying an intellectual (rational) intuition that can attain it. All this we can

21. Ibid., pp. 9, 36, 41–42.

safely say by confining ourselves to Kant's views on knowledge, setting aside for the moment his views on morality and belief. On the other hand, the contrast of a (cognitively ineffable) thing in itself with a (cognitively accessible) phenomenal realm is central to everything Kant has to say about knowledge. In fact, that contrast is even central to his repudiation of *knowable* noumena.[22] What Kant is actually saying, and it is a subtle point, is that if we possessed the power of intellectual intuition, as we do not, the thing in itself would indeed be noumenal for us: we should be able to know it without a formative/constitutive contribution of language or mind.

It is a troublesome circle of ideas to be sure. But what generates the trouble is something Putnam himself accepts: the notion that knowing is indeed a forming, making, or constituting, but not a forming, making, or constituting that totally creates the thing known. So he must also accept that part of the circle which includes the (positively empty) notion of the thing in itself. Thus, when Putnam says that we have no right to assign any property to the thing in itself (not even a strict dispositional property), he makes no point against Kant, for that is precisely what Kant says.

For Kant the doctrine of the thing in itself means only that our understanding and sensibility form in their union an objective (phenomenal) world, and that they do so not by virtue of their formative power alone but rather by virtue of the response of that power to something distinct from it. Does Putnam suppose that what *he* calls the real (rather than the phenomenal) is wrought entirely by language or mind? Certainly his own linguistic-conceptual relativity requires no such doctrine, and if it did it would collapse into "sheer linguistic idealism," which he himself repudiates. There are various other observations in Putnam's text that make him closer to Kant on the matter of the thing in itself than he supposes.

22. The expression 'knowable noumena' is, of course, a pleonasm; I introduce it to make more emphatic the point that 'noumenon' and 'thing in itself' are not interchangeable expressions for Kant.

What Kant calls the thing in itself is merely the obverse of the recognition on the part of antirealists like Putnam that there is something, something distinct from language-cum-theory, that will not tolerate our saying anything merely arbitrary; as Putnam himself insists, it is not true that "anything goes." For any nonrealist who shrinks from idealism, what Kant calls the thing in itself is merely that in nature which, if we said anything arbitrary or otherwise wrongheaded, "would show us our mistake." If it should be true, as Putnam says, that "our conceptual scheme restricts the 'space' of descriptions available to us" but "does not predetermine the answers to our questions," then the thing in itself is whatever else contributes to determining the answers to our questions.[23] In short, as the reader will have concluded already, the principle of minimal realism, which is accepted by every antirealist who is not also an idealist, is equivalent to the Kantian doctrine of the thing in itself.

The reader will, I hope, forgive me if I insist that I am not arguing for the soundness of the notion of the thing in itself. I am only arguing that the doctrine of the thing in itself goes with the Kantian territory *as that territory is* and also with as much of it as Putnam has been able to survey and agree with. The real difficulty about the Kantian doctrine is what antirealism shares with it: the notion that whatever we know is formed, constituted, or made by us as knowers out of something that we have not formed, constituted, or made. This notion of the thing in itself is not banished by refraining from trying to circumvent the concomitant prohibition against attributing properties to the thing in itself.[24]

All this still leaves us with a fundamental question about Putnam, even though the reader might now be prepared to

23. Putnam, *Many Faces*, pp. 31, 32, 39.
24. According to Kant, we cannot satisfy *cognitively* the metaphysical nisus that manifests itself in our tendency to treat this ineffable reality/thing in itself as though it were a noumenon (or noumena) accessible to an intellectual intuition. The latter part of the first critique and parts of the *Prolegomena* are concerned with this nisus; the second critique and the *Grundlegung* propose a satisfaction of the nisus in *belief* grounded in moral action. That Putnam finds this feature of Kant's ethical views displeasing has nothing to do with the point I make here.

concede (a) that Putnam is a reasonably standard antirealist, or nonrealist; and (b) that he and Kant are not so far apart on the thing in itself as he supposes. The remaining question is whether Putnam is really arguing for either a strict Kantian or a flexible Kantian loading of experience. Presumably we may dismiss the former possibility at once, for although a strict Kantian loading might be consistent with some of what Putnam says about science, it can hardly be consistent with the conceptual relativity that now leads him to give equal weight to other human activities. A flexible Kantian loading, though it would bring with it its own difficulties, is at least consistent with linguistic-conceptual relativity. But the examples Putnam introduces to persuade us of conceptual relativity provide no account of a genuine permeation of *experience* by the formative power of language-cum-theory. Nothing remotely like a Kantian union of the categorial/conceptual with sensibility can be found in Putnam's examples. Like Goodman, he draws his crucial examples from logic and mathematics.[25] The soundness of his observations about these examples is not at issue. What is important is that the questions raised by the examples can be settled entirely by an examination of the languages used to propound them. We may be prepared to agree that some sort of conceptual relativity operates in these regions, but we shall probably not agree that a flexible Kantian loading of experience has been even proposed, let alone demonstrated.

Antirealists, despite their frequent invocation of Kant's name, are neither strict Kantians nor flexible Kantians: their actual position, unacknowledged to themselves and so never clearly articulated, is that any particular body of language-cum-theory exercises a merely linguistic-cum-theoretic "loading" of experience—that is, a "loading" in which someone

---

25. In the first example, he examines the relativity of the concept 'object' in two logical worlds, one propounded by Carnap and the other by an imagined Polish logician who is using a version of Lezniewski's mereology. The second example concerns the ontological status of the Euclidean plane: are the points in the plane parts of it or mere limits? Putnam concludes that "which entities are 'abstract entities' and which are 'concrete objects', at least, is version-relative"; Putnam, *Many Faces*, p. 19.

who holds a body of language-cum-theory (hereafter, in this immediate context, 'theory') *interprets* experience in accordance with what the theory requires, while the general form of the experience undergone remains obdurately what it would have been before the theory was held. It is important, first, that in a "loading" of this kind two different things are attended to—on the one hand, theory, on the other, experience; second, that what is experienced is *interpreted as,* or *understood to be,* what the theory prescribes; third, that the focus of attention may oscillate between experience and theory; and fourth, that one of the two may be preferred to the other, so that the other either goes unregarded or is deliberately disregarded. It is also important that what is here called experience is commonsense, or prima facie, experience—something common and public in its general texture; something about which agreement can be reached if conditions can be controlled within reason; something ineluctable enough as to be set beyond utter transformation by the holding of a body of language-cum-theory; something that undeceives us in due course, even if for a while we have permitted ourselves to be deceived about it. It does not matter for the moment just what it owes its ineluctable and public character to. It matters only that, whatever the reasons for its stability, commonsense experience is the arena in which science conducts the empirical part of its business. It is the arena whose stability makes possible the repeatability of experiments by other experimenters who do not hold the theory. And that is to say that the very public nature of science depends upon it.

Thus, if the evidence for the presence of the Higgs boson should emerge from experiments at one of the two (at this writing) competing particle colliders, and if one of the successful physicists should announce, "We have seen the Higgs boson," that would be an exaggeration, though a pardonable one. What would have in fact been seen is the cluster of complex instrumentation that makes up the detector where the beams collide, together with the record on a photographic plate, computer screen, or some other instrumental complex.

On the basis of a body of theory and what is seen, the event is then *interpreted as* the collision of two particle beams and the consequent destruction of particles and the creation of a spray of new particles, including the elusive boson. One thing we can be sure of: those numbers and marks are in principle accessible not only to the experts but also to any reasonably intelligent and observant person who is quite ignorant of the theory, completely disinterested about the competition between the two groups, and quite unable to make the automatic interpretation the experts have made. Indeed, if in principle the outcome cannot be determined by having this accurate nonscientist pass along the instrumental data (such and such numbers on the screens, such and such an angle between divergent lines of such and such a length) to the experts, we should have good grounds for doubting the experiment. We should suppose not that the scientist enjoyed a mode of experience that had been subtly transformed in a flexible Kantian way by the holding of a theory, but rather that the scientist, oscillating between attention to the theory and its fate and attention to the crucial meter or mark, had opted for the former.[26] Science rests, despite all the propaganda to the contrary, on a distinction between theory holding on the one hand and the enjoyment or undergoing of experience on the other.

It seems, then, that there is something spurious about the Kantianism (with respect to science) of consensus antirealists. They tell us that experience is loaded with the antirealities, irrealities, worlds/true versions, or propositional realities they suppose us to have made. But in practice they deal with these things not as items that are generative, in a strict Kantian or a flexible Kantian sense, of counterpart experiential worlds but rather as items of language-cum-theory. If worlds are involved, they are counterpart *entia rationis:* beings of reason entertained by theorists in their theoretic

---

26. Compare Thomas S. Kuhn, *The Structure of Scientific Revolutions*, 2d ed. (Chicago: University of Chicago Press, 1970), pp. 193–98.

occasions as they use these items of language, not experiential worlds in which we live with our bodies as well as our minds. "Worlds" of that kind can indeed dominate and control our commonsense experience: those whose lives are given over to theories of one kind or another may dismiss the findings of common sense as primitive or in some other sense unimportant; scientists may report commonsense experiences as though they were experiences of what theory interprets them to mean. But no antirealist philosopher seems to have even attempted to show that such "worlds" are experiential in the sense of either a strict or a flexible Kantian loading.

Nothing that has been said in this section about the importance of common sense should be taken as a denial that in the attending to a work of art we undergo, or enjoy, a mode of experience distinct from that of common sense, a mode of experience which makes a profound reality claim on us. Nonetheless, the experiential component of a work of art does not simply replace the commonsense mode of experience but rather exists in one sense beside it and in another sense within it.[27]

27. But who, except their own disciples, ever needed the permission of analytic philosophers to take this aspect of art seriously? The youthful philosophy of art of many of today's influential writers of the linguistic consensus came to them through positivism: art, though perhaps not nonsense like metaphysics, had only the kind of therapeutic emotive value I. A. Richards envisioned in *Science and Poetry* (London: Kegan Paul, Trench, & Trubner, 1926). These philosophers have at long last awakened from their dogmatic slumber. But it should be said roundly that they brought no news to anyone who already possessed a solid knowledge of Kant's influence on art theory. The great romantics, for instance, despite excesses engendered by their involvement in the idealist response to Kant, made a detailed case for art as both a mode of experience and a way of access to reality, and they made it more adequately and audaciously than recent analytic philosophers have done. Richards himself came to realize this well before professional analytic philosophers began to grasp the point that the writ of rationality runs well beyond the sciences: his second edition of *Science and Poetry* (1935) took back much of what had been said in the first; and his conversion was completed in *Coleridge on Imagination* (London: Kegan Paul, Trench, & Trubner, 1934) and *Speculative Instruments* (London: Routledge & Kegan Paul, 1955).

6 Why Antirealist Relativism Creates No Problem for Science but Many Problems for Other Fields

Science uses, in the long run, an effective criterion for abandoning one body of language-cum-theory for another, or one subtheory for another: anticipation and control of experience, and prima facie experience at that. It is true that in certain parts of science, especially the physics of the very small and evanescent and the very large and remote, there may often be several competing theories that meet the criterion with the same degree of adequacy. That is a harmless enough relativism. For the most part, it depends on a temporary impasse—sometimes in instrumentation, sometimes in the complex reasoning that is designed to establish an appropriate connection between a certain body of language-cum-theory and some test or tests that can be assessed in terms of prima facie experience, and sometimes in the structure of the competing theories themselves. It is part of the telos of science that competing theories should in due course be resolved into one. From this perspective, it is unsatisfactory that, for all the talk of a grand unified theory, there are many candidates for that honor. A multiplicity of such theories might, of course, be the sign of a permanent impasse. But that too would be a harmless relativism, as long as the same experiential arena was respected by all theorists. The same thing can be said of the complementarity principle, which many suppose to be a permanent impasse. Relativism in science becomes troublesome only when scientists begin to suppose that impasses of this kind, whether temporary or permanent, authorize them to disregard the check of prima facie experience and construct their own realities. It is rare that scientists even *seem* to be in the grip of that supposition in their working lives. Scientists who philosophize are sometimes in the grip of it, but if we look at the working scientific careers of such scientists, we find a respect for prima facie experience as strong as that of any working scientist who has never been distracted by epistemology.

On the other hand, the consequences for the many other fields in which human rationality is deployed are profound and deleterious. In today's *Kulturkampf* the metaphor of language is everywhere, and in the human sciences, history, criticism of the arts, religion, and ethics it is increasingly used to relativistic effect.[28] This is not an essay in cultural history, so I do not mean to examine this nonscientific relativism as a cultural phenomenon. To those who are interested in combating it, however, the conclusions reached in this chapter offer some hope: linguistic relativism is not the irremediable condition it would have been if it were indeed the result of a flexible Kantian loading of experience. There is good reason to believe that the antirealities, irrealities, appearances, "worlds," or propositional realities linguistic relativists offer us to account for what we do and what we encounter when we are engaged in a nonscientific discipline have no experiential significance at all. On closer scrutiny they turn out to be no more than propositional structures, conceptual structures, theories, *entia rationis:* things ingeniously contrived by the formative power of our rationality and then entertained and argued about. Since there is no general acceptance of an empirical check on such activity analogous to the de facto empirical check on science that commonsense experience provides, one propositional complex can only be checked by another. Moreover, those who carry out this activity—it may be roughly described as argument, conversation, or dialectic—have so given over their lives to it that it never occurs to them that there is a domain of experience, one quite acces-

28. One field in which linguistic relativism has been notably successful is psychotherapy. A typical case is the influence on therapists of the antirealistic doctrine of the biologist Umberto Maturana. With a colleague who is herself a therapist, I have recently examined the influence of his work on that field; Barbara S. Held and Edward Pols, "Dell on Maturana: A Real Foundation for Family Therapy?" *Psychotherapy* 24, 3S (1987): 455–61. The article is followed in the same issue by a response from Dell and by our reply to him, pp. 462–68. For an earlier and more comprehensive article about the influence of the realism-antirealism debate on therapy, see Held and Pols, "The Confusion about Epistemology and 'Epistemology'—and What to Do about It," *Family Process* 24 (1985): 509–17.

sible to us as philosophers, that is not a Kantian product of one of those propositional structures and so does not confirm by *its* inner coherence whatever coherence that propositional structure might have. Thus they carry on the conversation of academic mankind and suppose it to be the conversation of mankind in general. That it should be neither conclusive nor progressive should surprise no one. The cultural world is swiftly becoming a congeries of linguistic communities, each providing a linguistic "reality" that purports to have an experiential counterpart but is in truth entirely self-sustaining. These communities then function as power centers that make self-fulfilling the contention of some of them that what purports to be the pursuit of truth is in fact the pursuit of power.

Among those linguistic communities there is none so influential as that of the linguistic consensus itself—the very propositional structure whose ramifications we have been examining. If we take the dogmas that define the consensus and put them together with the views of the consensus on the realism-antirealism issue, we find that the whole of the resulting propositional complex imposes a merely linguistic-theoretic "loading" of experience. In this case, the experience that is "loaded," that is, conceptually or propositionally interpreted, is not only commonsense experience but also the more reflexive development of that mode of experience—one in which we are rationally aware of ourselves as cognitive beings in a world our rational powers have had no hand in making. But it is truly a mere conceptual interpretation that the linguistic consensus offers us. Fascinated by this propositional structure, philosophers attend to *it*, even though all around it and around themselves there stirs, unaltered and always available if they but turned their attention towards it, a world they have not made. But to do that, they must work at perfecting a reflexive understanding of the radically realistic power of our rational awareness in its engagement with that world.

5

- Radical Realism:
  The Venue of Direct Knowing

1 *Two Functions of Rationality: The Function of Rational Awareness and the Formative Function*

- In the jurisdiction to which we now remove the case of realism versus antirealism, the rights of experience are restored: it is now seen to be a consequence of rationality's experiential engagement with the *real*. By the same change of venue the overweening claims of theory and language are restrained: they are now seen to be indeed creatures of the formative power of rationality, but to be destined for other tasks than the informing of experience. We make the move not by propounding a theory about how our rationality operates but rather by a linguistic gesture that originates in the same rational-experiential engagement that is its theme. The gesture merely calls attention to the two most basic functions of our rationality and gives them names.

By virtue of the function of rational awareness we *know;* by virtue of the formative function of rationality we *make*. The two functions are intimately related, and they cooperate in our most important enterprises. Thus, the formative function is responsible for the form of such things as language, theory, and physical artifacts; and all these things are intricately involved with our knowing, and the more intricately involved as knowledge grows more ample and profound. That involvement, however, is different from anything envisioned by either wing of the linguistic consensus. The formative function is not responsible for the form of such things as trees and persons, although it would indeed be responsible if the ontological status of such things were adequately expressed by saying that they belong to the ontology of commonsense language, or that they are postulated by the use of that language or of folk theory embedded in it. The two functions are distinct, and their distinctness must be kept before us if we wish to avoid misunderstanding the birthright of our rationality. By virtue of one function, we become rationally aware of such things as trees, persons, and other temporospatial things that rationality has not made but that come within its experiential range. By virtue of it we also become aware of theories, linguistic items, and other artifacts, and become aware of them just as the other, or formative, function of reason has made them to be. In neither case does rationality bring a formative power into the cognitive transaction: it becomes aware of the first group of things in their non-rationally made condition and of the second group in their rationally made condition.

As we exercise the function of rational awareness, our experience of the thing or things known enters into and sustains the engagement of our rationality with them. The awareness we are considering is thus not merely experiential and so does not stand in contrast with rationality. It is *rationality* that is experientially engaged with what it knows; so the awareness is rational-experiential, and the two aspects of reason and experience are inseparable. The function, act, state, or condition here called rational awareness may therefore be taken as the paradigm of the function, act, state, or condition of knowing. The character of the awareness is qualified, on the one hand, by all the universality we associate with rationality and have traditionally discussed in terms of ideas, concepts, forms, universals, and propositions; and on the other hand, by all the particularity we associate with experience and have traditionally discussed in terms of the senses, the here and now, the relativity of the here, the evanescence of the now.

No product of the formative function of rationality is insinuated between the function of rational awareness and what it attends to. By virtue of rational awareness we directly engage natural things like trees and persons, physical artifacts like chairs, more complex artifacts like the printed page now before the reader, and a vast variety of other temporospatial things that come within our experiential range. Whatever we attend to, it is the thing in question rather than some idea, concept, or proposition about it that suffuses our rational awareness. Although that function actively *attains* what it attends to, it is neutrally receptive in the sense that it contributes nothing to the enterprise except its own functional activity of being rationally aware: it does not, in actualizing itself, *form* that in which it actualizes itself. Our rational response to the reality of things attended to is an experiencing of them, but an experiencing which, because it is a *rational* one, is not captured in any of the notions of experience that have come down to us from the empirical tradition. It is not an experience that is given *to* rationality to cope with, because it is an outcome of the activity of rationality; on the other hand, that activity does not provide the structure of what is experienced. The experience is an experience *on the part of* rationality, and in enjoying it rationality enjoys the real thing or things it is attending to. The involvement of rationality in the experiencing means that we do not merely enjoy severally the things experienced; indeed, the particularity of things cannot be *rationally* enjoyed and at the same time enjoyed merely severally. In responding severally to the real particularity of the things attended to, rational awareness thus responds also to the reality of the universal unity they share. And that U-factor, as I called it in Chapter 2, although truly a factor in the being of the several temporospatial things, is not itself a temporospatial thing. We shall be returning to that delicate issue in due course.

The person who now reads this is a part of the enterprise no less than the person who writes. The reasons for this are complex enough, but for the moment it is important that we are beginning with a prima facie mode of knowing that is familiar to both of us: that in which an experience of the thing or things known enters into and sustains what is also the engagement of our rationality. The two poles of experience and reason—distinguishable, but only as two inseparable features are distinguishable—together make up the function, act, state, or condition we call knowing. Our way of proceeding is therefore both interpersonal and reflexive: to make progress, we shall be drawing together on the same mode of attention I am trying to call attention to. We are trying to attend rationally to each of the functions, to persist in that attention until it develops into an adequate rational awareness, and then to labor to stabilize that awareness in adequate language and so acknowledge the profound difference between the two functions. In that difference, in that intimate cooperation lies the secret of our epistemic condition.

But our labor is primarily a reflexive one: our first function—rational awareness—comes to know itself and then comes to know the other function and its own cooperation with that other function. I have already begun to draw upon that reflexivity of rational awareness and to try to induce the reader to draw upon it as well. If the reader accepts my inducement, our common justification will consist in a further reflexive deployment and intensification of the same function. It will, I think, turn out that we exercise the function of rational awareness not only in the kind of instance just given but in every argumentative context as well. If that is so, justification of whatever kind depends upon rational awareness, since the more familiar justificatory procedures to be found in contemporary philosophy are all argumentative procedures, and indeed the case made out for them is also argumentative. Although the full development of all this still lies before us, it seems best to drop the qualification 'prima facie' when we are talking about the function. The reader will in any event be able to judge at the end whether the justification has been successful.

We are often rationally aware of things that are indeed products of the formative power of rationality—our houses and most of the things that fill them and indeed all the products of our technology, from the simple to the dauntingly complex. In an obvious sense, the formative function does contribute something to the ontological status of what we are then rationally aware of. But that does not tell against the radical realism we are now considering. Once again we need only notice that the two functions are distinct: we become rationally aware of our artifacts as artifacts, and as precisely these artifacts; we acknowledge them as rationally made; and in doing so we deploy only the function of rational awareness. If it were not for the existence of a quite different realm of artifacts, the realm of the propositional, and the alleged role of that realm in ruling out a rational awareness that is independent of the formative power of rationality, it would scarcely be necessary to make this point. We shall soon be turning to that realm, but for the moment we confine ourselves to rational awareness of temporospatial things, both natural and artificial.

2   *The Primary Mode of the Function of Rational Awareness*

There is something fundamental and primordial about our rational awareness of temporospatial things. All knowledge begins there, for it must have been a primitive exercise of rational awareness that led some early human beings to become rationally aware—well before they could utter a proposition to that effect—that the things about them and the operations of those things could be named. To mark this point, let us call our rational awareness of temporospatial things *primary rational awareness.*

As we deploy the function, perform the act, enjoy the state or condition, we trace the temporospatial existence of what we are attending to. Our awareness itself takes up time, and it is saturated with the temporality of what it is focused on; and it also takes in a spatial range that is integral with that temporality. But although tracing it, tracking it, following it, surveying it, living along with it, coexisting with it—as we do in the case of our attention to another person, a bird, a tree, or a chair—we are not compelled to survey the whole of its temporospatial range to allow the function to complete itself. A small part of that range suffices, and the interruption of our attention to that other being (as we perhaps attend for a moment to something else) seems to offer no obstacle to our rational attainment of it.

The reason the function can complete itself in a small part of the temporospatial scope of whatever it attends to is that it is the function of *rational* awareness. The tracing of the temporal career and spatial deployment of the thing is therefore not a matter of mere experiential receptivity: it is instead integral with the actualization of rationality *in* our awareness. That means that our awareness is a response not just to the particularity of some portion of an entity's temporospatial extensiveness but also to a feature of the entity that transcends that portion because it qualifies all such portions. It is the unity of the thing in question, which, as noted earlier, is a unity that cannot be adequately expressed as merely the unity of that particular. Fragment though that portion may be of the temporospatial totality of the particular thing attended to, our tracing of it therefore gives us not just what is particular to the thing but also all the experiential base we need to recognize the presence of the U-factor in the particular. And that factor is common not just to those temporospatial things that resemble the particular in question but to all temporospatial things, including ourselves as embodied knowers. Today philosophers tend to take it for granted that all we can mean by the universal, common, or general feature of any thing we attend to is its amenability to being brought under concepts, or, if not concepts (which according to some do not in fact exist) then at least general terms, which in turn owe their generality to the generality built into the propositional setting in which they have their being. If the rational awareness I am now trying to call attention to is real, then the U-factor in the things it attends to is what makes possible a propositional approach on our part. If so, we invert the natural order when we suppose that the realm of the propositional (linguistic) supplies the universality of form that makes both experience and knowledge possible.

The justification of this claim is once more reflexive, reflexive in a sense that is integral with the exercise of the function and thus integral with the attainment of whatever we are attending to. If we allow the function reflexive scope as it attends to what is before it, we shall find that the traditional demand that knowledge be characterized by universality is satisfied not by the imposition of something formal but rather by the recognizing of something universal as an aspect of what is attended to as well as an aspect of the rational function that attends. The reflexive deployment of the function—a deployment which does not take place after its engagement with particular things but is rather integral with that—thus seems to be vital to the full actuality of the function.

The achievement of rational awareness is accompanied by a joint satisfaction that is familiar but sui generis and thus difficult to characterize. We may call attention to its double nature by noting that it has both an experiential and a rational pole and that these poles are inseparable. I say 'experiential' rather than 'empirical' to prevent confusion with the parallel situation in science, in which the rational pole and the empirical pole are separable. In rational awareness we enjoy this joint satisfaction as we know other things—in the first instance, temporospatial things; and we enjoy it as we ultimately come to know ourselves in the reflexive deployment of rational awareness. But it is not like other satisfactions. Let me defer discussion of this theme for a while, however, and make instead the negative point that the satisfaction does not terminate in the proposition or propositions in which we might give it expression. Thus, I express something of what lies outside my window just now: "A light moist snow is falling; it clings along the upper surfaces of the dark branches of the oak outside my window." But my rationality has not waked in the first instance to those propositions, nor has it waked to something shaped by those propositions, but rather to those interrelated temporospatial beings, independent of itself and its formative powers, that lie outside the window. The formative power of rationality responds to the achievement of rational awareness by forming the propositions that now lie side by side with those temporospatial things.[1]

In virtue of those propositions the rational awareness is stabilized, retained, made communicable. The propositions constitute an acknowledgment of the things of which we are rationally aware, but they no more create the things (or our experience of the things) thus acknowledged than a nod creates the friend whose presence it acknowledges. Propositions are items made by the formative function in response to the function of rational awareness. The formative function responds to the findings of rational awareness by an act of mimesis: it does its best to imitate what satisfies us in that inimitable way in which the particular is integral with absolute universality. The resultant *made* items are at once particular (just this concept, this word) and universal (this concept, this word, applies to a whole range of particulars). In that sense, and in that alone, the second function completes the first: propositions are formed residues of the acts of rational awareness they record and stabilize. We can also be rationally aware of propositions as such, for they are artificial objects, though not temporospatial ones. And we can respond to them

---

1. The propositions can thus be used to represent those things. This is by no means to claim that knowing is in essence representational. It is the particular virtue of rational awareness that it is not representational. But if we are rationally aware of two kinds of things—temporospatial ones and propositions—there will certainly be occasions on which the latter, especially in cooperation with memory and our imaginative powers, can represent the former.

by forming still other propositions and propositional things. Much of our intellectual life consists in this—so much so that intellectuals often suppose it to be the whole of that life.

If primary rational awareness is not focused on propositions or on language-cum-theory in general, neither is it focused on such items as impressions, ideas, concepts, or universals. Whatever the ontological status of such items may be, they do not, by virtue of some privileged direct access to them on our part, mediate our rational-experiential engagement with temporospatial things and accordingly make that engagement indirect and philosophically problematic. If we can in fact become rationally aware of such items, that awareness follows upon and does not precede our rational awareness of temporospatial things. Let us say, then, that primary rational awareness is direct—direct in the sense that it does not depend on a prior awareness of quite different items that tradition has assigned sometimes to our empirical side and sometimes to our rational side.

Primary rational awareness is not, however, absolutely unmediated; it is merely cognitively unmediated. Cognitive directness is compatible with the mediation of knowing by the structure and physiology of the central nervous system—more generally, by the body of the knower. It suffices for the possibility of direct rational awareness of temporospatial things that the contribution of the nervous system not include either (a) items internal to the knower which the knower becomes directly rationally aware of and then uses to develop indirect knowledge of temporospatial things, or (b) stimuli of which the knower cannot in principle become rationally aware, but to which it can respond by forming other items (for instance, ideas, concepts, hypotheses, or other theoretic-linguistic structures) which it can then become rationally aware of and use to develop indirect knowledge of temporospatial things.

It is irrelevant to the consideration of the function of primary rational awareness that we may not yet be able to give an adequate account of its epistemological structure—that we cannot as yet say, for instance, what role percepts (or sensa), concepts, or ideas play in it, or even whether there really are such items. It is also irrelevant that we are not yet able to say in any detailed way how precisely the physiological infrastructure of the function works. It is the ontic level of the function itself that we are exercising and enjoying, and unless we are persuaded that the full concreteness of any prima facie function is in principle an illusion, one soon to be dispelled by philosophical or scientific understanding, we had best go on trying to let the function deploy itself reflexively and, in doing so, reveal itself to us.

Whatever the detailed story of its infrastructure may be, the primary rational awareness I am calling attention to is directly realized or actualized in an experience of the temporospatial thing or things attended to. We do not come to know such things indirectly, as things inferred from ideas or concepts, or as things postulated by our adoption of certain linguistic turns—things whose ontological status is exhaustively defined by saying that they belong to the ontology of that language. There are special circumstances in which certain temporospatial things are known in one of these indirect ways. If, for instance, there are indeed quarks in some sense more fundamental than that some theory postulates them—that is, if what is thus postulated (or at least something very like it) truly exists outside the theory—then it is in this indirect way that we know them to exist (know that the postulation is warranted). But things that can be known only in that way are of a radically different magnitude or duration from the knower and thus have a different status from the temporospatial things we have been using as examples so far. If, for instance, there are persons in some sense more concrete and fundamental than that common language postulates them, we do not know them indirectly. There will, of course, be many things in the infrastructures of persons that we cannot know directly, but that is a different matter. Primary rational awareness may therefore also be called *direct knowing of the primary kind.*

3 Three Scandals That Hinder Recognition of Primary Rational Awareness

To say that our primary rational awareness of temporospatial things is direct is to say only that the dipolar satisfaction we ascribe to it terminates in nothing internal to the knower, whether received or factitious, but rather in the temporospatial thing or things to which it is attending. Familiar as this satisfaction is, philosophical contemplation of it has provoked incredulity from the very beginnings of philosophy. Until recently it has not been rational awareness as such that has provoked incredulity, for under other names and concerning things internal to the knower it has been taken seriously enough both in the remote past and through much of the modern era. What has provoked incredulity and created scandal from the very beginning is the thought that rational awareness might terminate in and be satisfied in temporospatial beings. That thought, so unsettling to philosophers, comes from common sense, which, before the intervention of philosophy, supposes both that there are things "out there" and that we know them as they are in their independence, their discreteness, and their unity. The intervention of philosophy has created three scandals that have stood in the way of our taking seriously this apparent birthright of our experientially engaged rationality. Let us call them the scandal of the Platonists, the scandal of the Cartesians, and the scandal of the scientists.

The scandal of the Platonists was not our apparent enjoyment of a rational awareness of real things not of our own making, for the Platonists took that to be precisely the vocation of rationality. What scandalized them and what they accordingly repudiated was the thought that those real things should be temporospatial ones. The Platonic repudiation, however, was a creative one: in the most momentous of all philosophic failures to distinguish the function of rational awareness from the formative function of rationality, Plato invented Platonic forms and told his followers, in effect, that, when we are indeed rationally aware, these are the things of which we are rationally aware.[2]

Whatever else Platonic forms are—and to me they seem to be, at the least, well-founded *entia rationis*—they are close cousins to words, to propositions, and in the long run to bodies of language-cum-theory.[3] Philosophers, moving among them as though in so moving they moved within the real itself in an unqualified sense, learned much about the nature of rational discourse, and in due time much about the nature of theory. Nonetheless, they had in effect taken as their standard for the satisfaction of rational awareness precisely the kind of satisfaction rationality can take in an *ens rationis* made (though certainly not ex nihilo) by its own formative function. It is an irony that they supposed themselves, as they pursued their dialectic, to have become rationally aware of things whose precise virtue was that they were independent of— were by no means a product of—the formative power of rationality. In any event, in turning to Platonic forms they had repudiated the effort to move from commonsense, or prima facie, rational awareness to an adequate reflexive grasp of the nature of rational awareness itself.

The scandal of the Cartesians was the thought that either the commonsense stance or the Platonic one could give us an immediate rational satisfaction in the real, regarded as something independent of the attentive mind.[4] They supposed instead that, if the mind were directly aware of something, that thing could only be in, and part of, the mind

2. Knowledge of the form of the Good carries us beyond the dialectic of the other forms, according to Plato; one consequence of this is that we are never quite sure whether it is appropriate to call the Good a form.

3. It should be noted that Aristotelian essences have more things in common with Platonic forms than is readily apparent in most accounts of Aristotle.

4. The term 'Cartesian' is here used in the broadest sense: it takes in all epistemology, whether empiricist or rationalist, that is based in any way on Descartes's very generous notion of ideas; he counts many items as ideas that count as impressions in Hume's empiricism.

itself.[5] So it was that, for Descartes and those who followed him, only our own ideas—ideas, that is, that were ontologically dependent on the mind that entertained them—could be directly known. In the terminology of this book, Descartes supposed that rational awareness can only attend to that which has its being *in* rational awareness. All other knowledge is indirect in the sense that it must be inferred from the representative reality of ideas resident in, and ontologically an expression of, the thinking being; most of the use Descartes makes of the notion of the clearness and distinctness of ideas is in the interest of this movement from ideas to what they are said to represent.[6] The chief example of this movement is the tortuous argument for the existence of the external world in Meditation VI.

Although this is not the place to pursue the matter, it is safe to say that both of these scandals persist in contemporary philosophy—the scandal of the Cartesians being especially prominent. We must now add to them the scandal of the scientists. What scandalizes the scientific spirit is not just the

5. That is what Descartes meant by the actual, formal, or subjective reality of an idea. That sense of 'subjective' is the ultimate source of our modern sense of subjectivity—as indeed is Descartes's philosophy in general; but it is important to recall that for Descartes 'subjective reality' still meant 'reality *in* a subject', and that the subject, though in this case a mind, could be any real thing thought of as the ontological support of its characteristics or features. In short, 'subject' here is close to the Aristotelian sense of 'subject' in the *Categories*. So, if one speaks, in a Cartesian setting, of the subjective reality of an idea, one merely calls attention to the fact that the idea actually exists in a mind and is moreover ontologically dependent on the mind. In that sense of 'subjective', the hardness of a stone also has subjective reality: it is a characteristic that exists in the subject, the stone, on which it depends.

6. In Descartes's doctrine, the subjective (formal, actual) reality of an idea is contrasted with its representative reality, that is, what the idea represents (to the mind that entertains it) as existing independently of the mind. Sometimes, according to Descartes, what is represented as existing does in fact exist; sometimes it does not. Since for Descartes the expression 'representative reality' is interchangeable with the term 'objective reality' (object for the mind), it is clear enough that the way we understand the terms 'subjective' and 'objective' today is in many ways an inversion of the Cartesian sense. This is in no way in conflict with our tracing most modern subjectivity (in our sense) back to Descartes.

thought that we are directly and rationally aware of temporospatial things and a temporospatial world, but also the thought that we are directly and rationally aware of anything at all—a Platonic form, or a proposition, or even a Cartesian idea. But the whole of this scandal is difficult to deal with in a short space, so I content myself here with sketching the recoil of the scientific spirit from the thought that a so-called rational person *here*—that is, a temporospatial being here—is capable of becoming directly and rationally aware of some other temporospatial being *there,* at some little or great remove from itself.

It does not scandalize the scientific spirit to suppose that the two things, knower and known, are in causal concourse with one another—that the so-called person is stimulated by the other being and responds to it with appropriate behavior, linguistic or of some other kind. Nor is it deemed scandalous to suppose that there is some isomorphism between physical events in the central nervous system and other physical events outside it with which the central nervous system is in causal concourse. Although those features of the transaction are acknowledged to be peculiarly hard to understand, it seems feasible enough to the scientific attitude that they will eventually be brought under physical laws of the usual scientific kind. What seems, to the scientific spirit, utterly intractable to those methods is a supposed rational awareness that on the one hand is involved with the physical events in the nervous system and on the other hand transcends them by being a sui generis attainment of the other being. What scandalizes is that the one being *here* should truly have a rational awareness of the other being *there* as it independently is. To someone in the grip of the scientific attitude, it is as though the supposed rational awareness were not after all confined to its own spatiotemporal locus but inhabited also the other one. What scandalizes, then, is precisely the prima facie radically realistic birthright of our rational awareness. It is professionally all right to exercise our birthright constantly as we do in common sense; it is even all right to fall back on it when we need it to get a formal semantics under way. But to take it seriously in the sense of making it an integral part of philosophy done in a scientific spirit would be scandalous.

4 The Secondary Mode of the Function of Rational Awareness

I have begun with our rational awareness (direct knowing) of temporospatial entities because I think they are the most important and fundamental things we know. But there are many other instances of rational awareness in which the thing known is not temporospatial. Among these things propositions are of the first importance, for there is a sense in which any known body of theory or any known doctrine consists of propositions. I do not think for a moment that this means that a body of theory has the structure of a language; it is more accurate to say that a body of theory is expressed in a language.[7] For that matter, it is only part of the story about

---

7. The language in which a body of theory is expressed is always some natural language that has been suitably enriched. But we do not do justice to the formative function of rationality if we suppose that it merely creates the linguistic structures by virtue of which expression takes place; it also produces the very entities that enter into that expression. To put the matter another way, the very expression of a body of theory involves the production, formation, construction, or making of at least some of the items that belong to the body of theory. Some of the items thus produced are linguistic terms, for instance, such terms of art as 'molecule', 'electron', 'quark', 'superstring', 'infinitesimal', 'set', 'group', and 'tensor'. Others are the entities or situations imagined or otherwise held before the attentive mind as the supposed correlatives of such terms, and the long-standing view that this entertaining involves items like ideas and concepts probably has more to it than most members of the linguistic consensus would concede. Sometimes there are also real correlatives to whatever is held before the attentive mind—real in the sense that they are not made by rationality and do not have their existence *in* rationality: there may or may not be superstrings that are both extratheoretically and extramentally real. At a certain level of complexity, a body of theory may also include complex symbolic artifacts which, by a metaphorical extension of the term 'language', are called formal or artificial languages. But I think that this does little to persuade us that a body of theory is essentially a *linguistic* thing.

bodies of theory and doctrines to say that they consist of propositions, for it may well be true that they include such items as ideas, concepts, and universals in their structure. However that may be, propositions play a significant role in bodies of theory, and we can certainly know propositions directly in the sense that we can be rationally aware of them both as particular propositions and as instances of the propositional. Indeed, attending rationally to propositions is something that philosophers do most of the time when they go about their professional business. It is this, no doubt, that prevents them from noticing that attending to a proposition as such requires an intellectual effort—precisely the effort of disengaging the propositional from its intimate association with whatever nonpropositional thing or things we may also be attending to. This in turn prevents them from noticing that the act or function of attending to propositions does not differ, qua rational awareness, from the act or function of attending to temporospatial things; the difference lies rather in the objects attended to.

When we attend to a proposition, say, 'snow is white', and know it to *be* a proposition and moreover just this one, we also have an *experience* of it. This does not mean that propositions are empirical items: the whiteness of snow is indeed an empirical matter, but the entertainment of the proposition 'snow is white' is not the entertainment of an empirical item. Nonetheless, we do experience the proposition as a proposition and as just this one: we experience it by way of contrast with our experience of temporospatial things like snow lying in a field on a winter day, or like the physical shape of 'snow is white' written in ink on a page. Part of our rational grasp of the proposition consists of our experiencing it to be a different kind of item from that snow and that ink on the page.

Thus, the obverse of the fact that " 'snow is white' is true" adds nothing to 'snow is white' is the fact that 'snow is white' emerges from our rational awareness, on one or many occasions, that snow is truly white. The proposition does not, that is, insinuate itself between ourselves and the uninterpreted real; we rather produce it as the completion and stabilization of our grasp of the uninterpreted real. When we are in rational-experiential touch with a reality that is of our own scale, propositions have not made it available to us in an interpreted form. On the contrary, our rational-experiential awareness of the uninterpreted thing makes the appropriate proposition available to us. Interpretation takes place at the level of the production of a (propositional) body of theory designed to give us indirect knowledge of things of which we do not, and in some cases cannot, have direct rational awareness. When they are formed into this body of theory, which we then attend to instead of attending to the thing, they do present us with an interpretation of the real that acts as a substitute for it.

Despite all this, our rational grasp of propositions is direct enough. When we entertain a proposition, it is before the attentive mind, to use a phrase that is as unfashionable as the word 'entertain'; the proposition is grasped, in a rational-experiential sense, *as* a proposition and as just the one in question. Let us say that we have a *secondary rational awareness* of the proposition and that our grasp of it is an instance of *direct knowing of the secondary kind*. It will be helpful to think of our experience of it as parasitic on our experience of ordinary temporospatial things.

The term 'secondary' is not designed to call attention to any indirectness, any lack of vividness or immediacy, in our rational awareness of the proposition. It is meant to call our attention to the rational-experiential function, act, state, or condition in which the knower knows the proposition and to a complex acknowledgment, or recognition, which accompanies that direct knowing. It is a recognition, first, that the knower has participated in the making or forming of the proposition thus known, and, second, that some thing known by primary rational awareness was also essential to that forming or making. No matter that the uttering or the entertaining of a proposition signalizes the completing and stabilizing of our primary rational awareness of the white snow lying in the field. It nonetheless remains true that when we are rationally aware of the proposition 'snow is white', that is, when we know it directly for what it is, we know it for something our rationality has made—know it, in effect, as a well-founded *ens rationis*. We profoundly misconceive both our function of rational awareness and the formative function of our rationality if we suppose that the linguistic products of the latter inevitably provide the rationally assimilable structure of the things we know.

5 *Rational-Experiential Satisfaction and the Natural Reflexivity of Rational Awareness*

Whenever the function of rational awareness is deployed, it is qualified by a natural reflexivity that is at least incipient even in supposedly nonreflexive uses. This reflexivity can be systematically cultivated and intensified. Although rational awareness is never actual except when addressed to some object or objects, the intensification of its reflexive component marches with equal pace with the intensity of our concern with the object and its independence. So we can call attention to the function only by also calling attention to whatever the function happens to be actualized or completed in. We can make the same point in terms of the satisfaction that is the outcome of successful attention. When, in reflexivity, we take satisfaction by attending to our own primary and secondary rational awareness, that satisfaction is dependent on the satisfaction we take in whatever thing or things, distinct from ourselves, we attain, grasp, entertain, see, attend to, or are in the cognitive presence of. In other words, the satisfaction of the reflexive factor in rational awareness depends on our satisfaction in the thing or things, distinct from ourselves, that we are rationally aware of (know directly). Nonetheless, the satisfaction in things distinct from ourselves depends on the heightening of the reflexive factor. The very nature of the independence of the thing attended to does not emerge except as our own relation to the object becomes part of what we are attending to.

Thus we can never attend *merely* to our own consciousness, subjectivity, rational intuition, mentality, intentionality, or rational awareness as such. And we can never attend to anything as independent of our own formative powers without attending reflexively to ourselves in rational-experiential engagement with that object. On the one hand, our reflexive attention is successful only by being occupied also with the reality of whatever thing or things the function so variously named has managed to attain: it is our enjoyment of the attained reality that makes the function reflexively available to itself. On the other hand, the reflexivity makes possible our attention to both the radical independence of the object and the radical involvement of its particularity with the U-factor our rationality shares with it. Taking seriously the radical reality of the thing or things attained by the function, we are led to take seriously the rational-experiential capacity we have brought into play in attaining it. Exercising reflexively the rational-experiential capacity of which rational awareness is the actuality, we are led to acknowledge the radical reality of the object—that is, its radical independence of the formative capacity of which our rationality is also capable.

No doubt many of the expressions I have just used to draw attention to rational awareness are metaphorical, and no doubt none of them expresses precisely what the knowing subject is doing. This would be a radical defect if what they call attention to were inaccessible and merely alleged to be like those things from which the metaphors are drawn. But the function is not inaccessible; it is the most familiar thing in the world. The real difficulty is that, although ubiquitous, it is sui generis: there is nothing quite like it, and indeed it is what we must call into play whenever we are asked to consider an alleged likeness between two things. So the metaphors at best call our reflexive attention to what they do not quite express.

The rational-experiential satisfaction that accompanies the actualization of rational awareness (whether of the primary or the secondary mode) is the only warrant we have both for the directness of rational awareness and its attainment of a reality that is independent, qua known, of the formative function. But the qualification 'qua known' is of some importance, for primary rational awareness deals sometimes with natural things, which are in no sense dependent on the formative function, and sometimes with physical artifacts, which are products of the formative function. The point of the qualification is that in the latter case the formative function does not enter into and qualify our rational awareness of those made objects. Similarly, although secondary rational awareness deals with the propositional, which is also a product of the formative function, the formative function does not enter into our rational awareness of the formed realm of the propositional.

The warrant of rational-experiential satisfaction becomes available, however, only with the intensification of the naturally reflexive factor that is always present in any rational awareness. One reason for this is that any particular instance of rational awareness can in principle be mistaken— mistaken, that is, not about its attainment of *something* that is independent, qua known, of the formative function, but about whether the particular thing attended to is what it purports to be. In any ambiguous perceptual situation, for instance, we may be attending to something nonpropositional that is not a product of the formative function without being sure precisely what it is. What has the cat caught, there in the half-light of a misty dawn? Is it a mouse or a small bird? We may opt for the theory that it is a bird and may even form the proposition 'it's probably a bird', thus attending both to one dubious nonpropositional thing that is nonetheless something independent of our formative power and to another thing that is not dubious and that, though qua proposition formed by us, is not qua *known* as that proposition formed by our act of attending to it. But still, the warrant for what we are sure of is not just this particular complex of certainty and uncertainty, but rather what our intensified reflexivity, in this and many other such instances, has managed to reap by way of permanent satisfaction. To put the matter another way, reflexive satisfaction, although always bound up with some particular instance of rational awareness (and thus with some particular satisfaction), heightens and makes dominant the U-factor that is one component in any particular satisfaction.

The function of rational awareness always actualizes itself in something that is independent, qua known, of the formative function; satisfaction is accordingly taken both in the independent particularity of the thing attended to and in the universality that is integral with each particular. The particular may be a real but ambiguous physical thing like the supposed bird. It may also be something imagined, as when, all the while I attend to that ambiguous physical thing, I also attend to an imagined bird. If the bird turns out to be merely an imagined thing, as when the cat has in fact caught a mouse, I am then attending in imagination to some particular thing I have formed, though it is real enough, qua formed particular. Again, the particular might be a proposition—a proposition real enough qua formed proposition—about what I had supposed the cat to have caught. Through a survey of particulars of all sorts, a survey that includes some uncertainty and some plain error, the reflexive function of rational awareness takes permanent possession of the U-factor they all participate in and instantiate. It thus finds itself in a position to affirm its own in-principle capacity to attain that which, qua known, is no product of the formative function. The affirmation carries with it a concession that in no way annuls it: in some measure rational awareness is always at the mercy of mere particularity. Thus, my uncertainty about the ambiguous thing seen at dawn in no way abrogates the radically realistic birthright of rational awareness.

It is important that the recognition of the compresence of the two aspects of particularity and universality depends on the natural reflexivity of rational awareness itself. The successful actualization of rational awareness in the independence (qua known) of the thing attended to carries with it the actualization of rational awareness in its own independence (qua self-known in reflexivity) of the formative function of reality. The reflexive component in rational awareness thus repudiates, by virtue of the nature of the satisfaction that goes with it, what someone, attending only to the propositions by which I now labor to call attention to something nonpropositional, might take to be a velleity toward idealism. Rational awareness, exercised appropriately—which means exercised also with reflexivity—creates no realities, least of all its own reality as a function both particular and universal.

Anyone who so exercises the naturally reflexive function of rational awareness as to enjoy and acknowledge the correlative rational-experiential satisfaction will, I think, also acquiesce in my attempt to rehabilitate the word 'satisfaction'. In its recent employment in epistemology, this term refers to something that befalls propositions and related items. Its use in that setting is a metaphorical extension of its home use, but an extension designed to make its home use inauthentic—as though the latter were a questionable metaphorical extension of a supposed rigorous home use in the propositional setting. If we now rehabilitate and deepen its real home use, we do so by noticing that the function of rational awareness when *in potentia* is characterized by a need, a need to achieve an end that is sui generis. We can do no better than to call that need a need to satisfy itself in that which is independent of whatever exercises (and thus instantiates) the function. It is a need for that which is universal on the one hand and particular, immediate, and sensory on the other. So it is both a need for a temporospatial causal link with what we attend to and a need to exceed or transcend that sort of causal link. That transcendence is merely the exercise of the ontic level that constitutes the function as what it is: we see what is the case, and by doing so transcend the relativity which is our cause-followed-temporally-by-effect connection with the thing attended to—as well as the relativity we should indeed be victims of if our only response to that causal nexus were to call into play the *formative* function of rationality to cope with whatever comes to us by virtue of that nexus. The double satisfaction, together with the reflexive intensification of it that is necessary to know its rights and limits, is the only guarantee we have that we can in principle know what is the case. Those who propound arguments to the contrary must rely on a series of like satisfactions as they move through the steps of their arguments.

The failure of philosophy is evidently a failure in the reflexive component of the function of rational awareness. Less technically expressed, it is a failure in self-knowledge on the part of our rational awareness: it fails to become adequately aware of itself in its rational-experiential engagement with the various things it attends to. It therefore fails to distinguish itself from the formative function of rationality. This failure has little practical resonance, for the two functions remain distinct for most of our rational occasions despite anything philosophers have to say about the matter. If the function of rational awareness has a radically realistic birthright, philosophy cannot annul it; it can only make us misprize it, look for substitutes for it, fail to develop its full potential. The failure affects our rational composure and self-confidence, and because of that it hinders the further development of our rational powers in the direction of first philosophy. We know some real things, but failing to recognize that achievement, we also fail to actualize our power to deepen our engagement with the real.

6 *Dependence of the Formative Function on the Function of Rational Awareness*

By virtue of the function of rational awareness we attend to and attain a variety of particular beings and aspects of beings *just as they are*. But, limited as we are both by our own particularity and by the nature of our sensory modalities, we are unable to become rationally aware of a whole range of particular beings and, no doubt, of a whole range of aspects of being as well. Even our rational awareness of beings that are indeed accessible to us—other persons, trees, chairs—is by no means exhaustive. Our rational awareness of other persons, for instance, may more or less adequately take in the ontic level of the person and several other levels—that of physical objects of a certain size, for instance—and still leave much of the infrastructure of the person unexplored. As for the universality of Being, although we become rationally aware of it just by virtue of the presence of the U-factor in each particular temporospatial thing we become rationally aware of, its superabundance is surely only vaguely known at this point in our development.

On the other hand, our attainment of what we do become rationally aware of is authentic as far as it goes. We are not, that is, at the mercy of the formative function of reason. The formative function does not qualify the function of rational awareness, and so it does not doom us to contemplate only what are in effect our own constructs. To suppose the contrary is to fail to notice how dependent the formative function of rationality is on the function of rational awareness. What the formative function makes, it makes not ex nihilo but rather out of what is available to it by virtue of rational awareness.

Let us rehearse some of the things the formative function does indeed make. First of all, there are our physical artifacts: houses and all the things that fill them, together with all the complex products of technology that make the more commonplace artifacts possible. Then there are all our works of art, some of which are physical artifacts as well, even if they are not adequately expressed by that category alone. But we shall not get very far in talking about the complex world of art without noticing that it includes many merely imagined things as well. These too are products of the formative function of rationality, though our formative powers in general no doubt rely also on a dynamism of feeling and sensibility not completely captured in the term 'rationality'. At any rate, the imagined things we might include under the rubric 'art' we might not wish to include without qualification under the rubric *entia rationis*. In any event things imagined—and thus in some sense products of the formative power—are much more extensive than the world of art.

More to our purpose just now are a host of other products of the formative function that are clearly rational: bodies of scientific language-cum-theory, together with all other things that are in any sense linguistic. We shall want to include in the latter group both natural and formal languages and all the ingredients of language as well. And we shall do well to remember that those ingredients include both formed physical things and the symbolic use those things are put to. Perhaps much of all this can be conveniently summed up under the rubric 'the propositional', provided that we remember that terminological debates about 'proposition', 'sentence', and 'statement' are by no means merely terminological. It is, in any event, one central thesis of this book that when our formative power produces the propositional, it does not do so ex nihilo, nor in response to stimuli that are either preexperiential or prerational, or both, but rather in response to our rational awareness of the nonpropositional. That the production of the world of the propositional enables us to focus sharply, to stabilize, and to recall our acts of rational awareness is beyond question. But in this section I have tried to make salient another matter, namely, that the production of the world of the propositional, a world based upon and now appearing beside and in contradistinction to the world of the nonpropositional, makes available to the function of rational awareness a whole new realm of beings—a realm of well-founded *entia rationis*.

7 Indirect Knowledge and the Formative Function

Bodies of theory, which are products of the formative function, have the power of providing us indirect knowledge, in certain circumstances, of a remarkable range of things not accessible to rational awareness—that is, things that cannot be directly known. It is the dependence of the formative function on rational awareness that in great part accounts for that power. What is made by the formative function is made out of what the function of rational awareness supplies to it, and so the radically realistic birthright of direct knowing is passed on to the formative power, giving the latter a reach and authority that is extraordinary even in circumstances in which we should be reluctant to call the outcome indirect *knowledge*.

Bodies of theory are formed in the first instance on the basis of things and features of the commonsense world that come within the range of the function of rational awareness. But the formative function also responds creatively to its own earlier by-products, so that in the gradual evolution of a body of theory a new theory is formed in part by way of response to older theories. What is usually overlooked, certainly by the linguistic consensus, is that our response to older theories is by virtue of our rational awareness of them. Direct knowing thus enters in a variety of ways into whatever we are able to do with language-cum-theory.

Among the many things we can do with a body of theory, it will be helpful to distinguish two extremes: a successful theory in which it seems clear that we have achieved indirect knowledge of certain entities that we cannot in principle know directly; another theory, successful in a qualified sense, in which what purports to be indirect knowledge of something we want to know directly turns out to be deceptive. The partial success of bodies of theory of the latter kind, I suggest, rests nonetheless on their capacity to provide indirect knowledge of certain general features of things that are relevant to the structure of the things we want to know.

The first extreme is illustrated by the body of theory dealing with the structure of DNA. The evidence now seems overwhelming that the familiar double helix made of plastic or some other substance is at least roughly isomorphic with a biological entity—call it the target structure—that exists in the cell nucleus. The target structure in this case is not directly known by us, and in principle it cannot be directly known. The study of DNA is an intricate business, and it is certainly not exhausted even in an elegant and relatively complete physical model constructed on a macroscopic scale. But rational awareness (direct knowledge) of such a structure appears to be rational awareness of something that is structurally analogous to the target structure. If so, direct knowledge of one thing is indirect knowledge of another. The argument for the isomorphism between the two structures is so complex that we must concede that our knowledge that rational awareness of the model constitutes indirect rational awareness of the target structure is itself indirect. On the other hand, that argument itself includes many instances of (direct) rational awareness: that of the macroscopic physical model, that of the x-ray crystallographs that help support the isomorphism, but also—for we must not forget the secondary mode of rational awareness by which we know each item of the body of theory qua item of theory—any element of the theory that cannot be modeled in physical terms. Indeed, if the general thesis of this book is sound, any step in any argument must itself be an act of (direct) rational awareness. The case of DNA is by no means unique, and the emphasis on physical models is no doubt already old-fashioned. Any working organic chemist knows how that discipline has been transformed by computer modeling. That those models permit us to become rationally aware of isomorphs of organic molecules that exist as extra-theoretic structures and so to become rationally aware, indirectly, of important features of the latter seems beyond question. Only the influence of the dogmatic claim that we cannot be rationally aware of anything that is not somehow formed by the intervention of language-cum-theory makes us think otherwise.

At the other extreme we have the case of older bodies of astronomical theory. The Aristotelian and the Ptolemaic bodies of theory have notable deficiencies, but they had considerable predictive success in their times. As in the case of DNA, physical models played an important role in those theories. We may reasonably suppose that first imagined structures and then physical instantiations of them were produced by the formative power of human rationality, and produced moreover in response to observations made, in the primary mode of rational awareness, within the commonsense world. That these models fail to provide indirect knowledge of their target structure in the sense in which the macroscopic plastic double helix does indeed provide indirect knowledge of its target structure seems clear enough. For the Aristotelian case, for instance, there are no crystalline spheres in precisely the same sense of 'are' in which there are indeed minuscule double helixes within the cell nucleus. Yet in their two ways the Aristotelian and Ptolemaic bodies of theory-cum-language do indeed save the phenomena, or at least some of them, and those phenomena belong to the commonsense world about which the theories afford predictions. So we may suppose—once again by an argument of some complexity, and so indirectly—some systematic relatedness between those antique models and some measurable properties of the real heavenly bodies.

If we now consider the two extremes as total bodies of theory (including the physical models) in terms of their relevance to the realism-antirealism debate, what may we conclude? It seems tolerably clear that contemporary DNA theory must be given a realist interpretation: if we confine ourselves to certain structural features of an entity within the cell nucleus, the theory seems to represent those features as they really—that is, extratheoretically—are. We of course set aside for the moment all sorts of dynamic features of the entities the structure comprises. It would appear that some of these features may be known indirectly by other portions of the background body of physical theory and that other features are known only in the sense that macroscopic phenomena associated with them are saved by the theory. But for the moment that is not our concern; we confine ourselves to features of the target entity that do seem to be isomorphic with the model.

The case is quite different with the other extreme, which for the moment we may take to be just the Aristotelian model of the heavens. The model, although intended to do more than save the phenomena, does only that: the system of heavenly bodies does not really consist of nested crystalline spheres, although supposing it to do so had its uses in its time. There are no such spheres in the same sense of 'are' in which there are indeed component units of DNA linked in the structure of the double helix. The interpretation of the whole body of Aristotelian theory must therefore be an antirealist one. In the sense of 'ontology' now current, ontology of a language, the crystalline spheres belong to the ontology of the body of language-cum-theory: they "are" for that body of language-cum-theory, but they have no other (extratheoretic) ontological status. They are unreal (in the sense of radical realism),[8] but if we wish to work with that theory we must assume them to be real.

To express oneself in this way is to fly in the face of assumptions that the realist and the antirealist wings of the linguistic consensus have in common. But the venue we have moved the dispute to prescribes a radically realist standard: the direct knowing of (some) of the transtheoretic, translinguistic real by way of the function of rational awareness. Whether we are dealing with a theory that can plausibly be given a realist interpretation or with one that seems to require an antirealist one, we are assuming a movement between real temporospatial things (known by the primary mode of rational awareness) and real bodies of language-cum-theory (made by the formative function of rationality and known for what they are by the secondary mode of rational awareness). In that sense the antirealist interpretation

---

8. This is not inconsistent with the point made in the previous paragraph, for there may be some systematic relatedness between those antique models (real qua models) and some real measurable properties of the (real) heavenly bodies.

I have just made about certain once useful bodies of theory paradoxically rests on the radical realism that dominates the judicial process of our present venue.

In dealing with the case of DNA, I assumed that its realist status had to be justified by a complex argument. The body of theory could be known by rational awareness, so our knowledge of its actual structure (qua formed body of theory) was direct. This carried with it the assurance that the body of theory had the status of a transtheoretic entity in the sense that our access to it was not by the mediation of another body of theory. Although many observations of temporospatial things comprised in that complex argument are also known directly, the structure of DNA itself cannot be known directly: its reality status is known by inference. But there are certainly some bodies of theory today whose realist interpretation does not need to be justified by a complex inference. One important case is the current body of theory about the solar system, for technology has vastly increased our power of rational awareness in the primary mode. What travelers to the moon have seen, together with what travelers will see who some day go to where our Jupiter probe has already gone, gives us solid nonargumentative grounds for the reality claims of the physical models that form part of that theory. The target structures, once inaccessible to our power of direct knowing, are now within our experiential range.

6

- Nine Theses about Science, Common Sense, and First Philosophy

1 *Introduction*

- The nine theses I put before the reader in this chapter sum up the consequences of radical realism for common sense, science, and first philosophy. The reader should not, however, regard the theses as a set of propositions to be considered in isolation from their nonpropositional origins and then assented to or rejected. I have drawn the theses out of the preceding five chapters, reviewing their leading themes and bringing their conclusions to a sharper focus. But because much of the material in those chapters was intended as an active exercise in rational awareness—an exercise beginning with the attainment by rational awareness of the nonlinguistic real and ending with its reflexive confirmation of that attainment—much in the theses may be regarded as the propositional outcome of that exercise. Radical realism is in all ways at odds with the dogmas of the linguistic consensus—above all, with the doctrine that our knowing terminates in propositions rather than in that which justifies the assertion of the propositions.

2   The Nine Theses

We begin with a thesis that goes back to the primary concern of first philosophy: the real, or Being, considered as something we can attend to and articulate only to the degree that we can also attend to and articulate the relation between our rationality and the real.

THESIS I. The primary and indispensable function of reason is rational awareness, which is direct in the limited sense that it does not depend upon our knowing—or for that matter merely experiencing—any intermediaries. By virtue of rational awareness we attain, in all their independence of the function or act that attains them, both particular things and the U-factor present in each—the unity and universality of reality, or Being. Although our attainment of the real is not exhaustive, being instead selective in ways that depend upon our finitude, of which the limitations of our sensory modalities provide the most salient example, we do not necessarily interpose cognitive, linguistic, or merely experiential intermediaries between ourselves and what we wish to attend to; nor do we in any sense form, by virtue of the act of attending, what we become rationally aware of. It is first philosophy, which is, before it can be anything else, rational awareness become reflexive, that tells us this, giving us in that way a universal assurance and self-confidence that allows us to acknowledge and overcome the shortcomings of any particular act of rational attention. Rational awareness is capable of focusing both on what we have in no sense made and on what we have indeed had a hand in making (including, but by no means limited to, the propositional); this flexibility of focus is the source of much of the confusion in epistemology.

The term 'attainment' is replaceable by others, for it merely calls reflexive attention to what is sui generis and familiar—an epiphany so commonplace and habitual that we overlook it or discount it. The indispensable feature of first philosophy is a reflexive epiphany in which both the object or objects of the commonplace epiphany and the commonplace epiphany itself are attained together: the reality that is the rational function we exercise in doing so now becomes part of what we attend to and attain. The term 'attainment' is meant to include the satisfaction that accompanies the successful actualization of the function in what it is directed toward.

Dominant epistemological doctrines, both that of the linguistic consensus and those of several other philosophies, tend to exclude the factor of awareness. They do so by focusing professional attention on propositional complexes that are the outcome, but by no means the sole or ultimate targets, of rational attention.[1] Although it is less plausible for philosophers to ignore or dismiss the factor of awareness when they are discussing the experience against which a body of theory is tested, it is nonetheless usual even in those circumstances to exclude the awareness that belongs to the knower and to concentrate instead on the propositional expression of some item experienced by the observer or experimenter. Awareness, however, is essential to all direct knowing, whether it is focused on some *ens rationis* or on some item belonging to the commonsense world. In direct knowing the articulation (in words and other symbols), the universality, and (sometimes) the a priori aspect appropriate to what I have been calling the rational pole concern *that of which the knower is aware* while the activity that achieves all this is taking place.

The articulation, important as it is, distracts us from what we were aiming at: though it is the whole point of the articulation to help us to reach what we were aiming at, to keep it in view, and to return to it at will, that goal is such a

1. Propositional complexes are here, as elsewhere in this book, regarded as *entia rationis*, although that category is meant to include many of our more important imaginative constructions as well.

difficult one that we tend to shift our attention to an easier and certainly more static target. Features that belong to the *entia rationis* we have formed—for instance, their static nature—are taken to be precisely what we had been aiming at. One thing we had indeed been aiming at, although we manage to learn this only after the reflexive turn that is the beginning of first philosophy, is the common unity of things—common because it is also the unity of *each* thing—that supports the generality we take for granted in articulate discourse and theory. To the degree that our glance wanders from from that U-factor and turns instead to the propositional structures that are the outcome of our engagement with it, we fall short of it. It is essential to notice that propositional structures work best when they so complement our engagement with their original that in our very use of them we take them to be poor substitutes for their original—when, that is, their *formed* status is salient all the while we are engaged with their original. One thing that helps distract us from this subtle and difficult task is the fact that propositional structures are also potential objects of our attention, and moreover important ones.

THESIS II. Knowing is, before all other things, an activity, function, state, or condition of the knower that completes itself in the independently real. All knowing has both a rational and an empirical, or experiential, pole; but in direct knowing—that is, rational awareness—the two poles are inseparably fused and in mutual support, even though each is distinguishable and partly characterizable. For that reason, it is best to avoid the usual overtones of the term 'empirical' and say that direct knowing has a rational and an experiential pole. But, in any event, the term 'awareness' is not meant to call attention to a mode of experience that functions at a subrational level, nor is the term 'rational' meant to call attention to some purely conceptual or propositional way of functioning. Rational awareness is *reason experiencing* rather than reason responding to experience.

Both the commonplace and the reflexive epiphany of Thesis I are dipolar in the sense of Thesis II. The rational pole is the pole in which the unity/universality which is the U-factor predominates, the experiential pole that of diversity and particularity. But the knower cannot attend to either pole in some pure form in which it is isolated from the other one. The fusion of the rational and the experiential poles in direct knowing means that the awareness supports and justifies the articulation, while the articulation focuses, intensifies, and stabilizes the awareness. Direct knowing can focus on the commonsense world (or some part of it) or on a body of scientific theory (or some part of it); in Chapter 5, we called these two focuses respectively primary rational engagement and secondary rational engagement.

The rational pole is a universal, symbol-generating transcendence of the here and now, but its functioning depends on the presence of the experiential pole—and the latter is unfailingly present, even in our most formal intellectual exercises. On the other hand, the symbol-generating feature of the rational pole depends on something that is also unfailingly present, the common unity of things. The *rationality* of our awareness always includes this U-factor, though at first only in a subliminal way, for we attend self-consciously to this factor only as the reflexive power of our direct knowing develops. But the fusion of the two poles must be taken seriously: although the rational pole plays the governing and active role in our cognitive access to the real, it does so only under the constraint of the experiential pole. The experiential pole, on the other hand, is our sense-based receptivity to the real by way of the particularity of the here and now, but only under the governance of the rational pole; so the receptivity is not merely an openness to particularity but rather an openness to reality by way of particularity.

THESIS III. The relation between the two poles is, however, different for science, and we mark this difference by a terminological change. The knowing we call science is a complex activity that swings cyclically between a rational and an *empirical* pole. Less sophisticated ways of knowing that prefigure science are to be found in our commonsense activities, and these are also cyclical. These cyclical ways of knowing depend on (direct) rational awareness in the sense that the knower is rationally aware not only of items belonging to the rational pole but also of items belonging to the empirical pole. The empirical pole, then, is something directly known rather than merely experienced.[2] When the cycle itself is successful, it affords us indirect knowledge of something we cannot now—and often in principle cannot ever—know directly. In the case of successful science, direct knowing of a body of theory provides indirect knowledge of what the theory is about.

The realism-antirealism controversy in philosophy of science has been generated mainly by the problem of the relation between a body of theory and what the theory is (in commonsense terms) about. That relation is no doubt full of difficulty, not least because some of the most important things a body of physical theory is about are not accessible to us directly. This is true of the smallest particles, which, it is often assumed, are of all things the most important for the scientific enterprise; it is even more true about the discrete quanta in which energy transformations take place. It is true about the very largest things as well—galaxies and clusters of galaxies, and the physical universe itself, considered as a totality in perhaps infinite expansion. It is true also of some things whose natures are incompatible with direct accessibility—black holes, for instance, if they do indeed have an extratheoretic existence.

2. The linguistic consensus would agree that the commonsense world is not *merely* experienced, for the consensus also agrees with the more general claim of radical realism that a pure or mere experience is not to be met with. It would not, however, agree with radical realism's claim that, by virtue of direct rational awareness, commonsense things are known to be in principle independent of the formative power of our rationality.

These difficulties should not be used to propound a general epistemological problem, for two related reasons. First, a great many things that we know are readily accessible to us, being either of the same temperospatial scale as ourselves or else in some sense made by us (as a body of theory is made by us). Second, such things are not known only by way of language-cum-theory, although a body of theory can indeed be brought to bear on some of them, and although such theory can erode our confidence in the adequacy of what purports to be our direct and nontheoretic confrontation with them. Our direct knowing of a mathematical model of a black hole in the context of a body of appropriate theory is our only way of knowing black holes indirectly. If there should indeed be black holes, they are real in some other sense then that in which the model is real. Everything near a black hole falls into it; nothing falls into a model. Meanwhile, the term 'black hole' has a merely theoretic status; despite all the weight of scientific opinion that says, in the realistic spirit of the working scientist, that there are extratheoretic black holes, it is possible that scientists will one day think otherwise.

The concept 'person'—or the term 'person'—has a different status, for there are persons directly before us to be seen, touched, and communicated with; and they are present to our rational awareness in such a way that we can scarcely say that we use the term 'person', imbedded in a certain linguistic context, because we lack direct access to a supposed original. There is an intimate relation between the term 'person' and what we are rationally aware of that is radically different from the relation 'black hole' bears to a black hole, even if there are real black holes. This is obscured for us by the fact that the *terms* 'person' and 'black hole'—both in some sense formed by our rationality—can be at the focus of our rational awareness in precisely the same way. And of course many philosophers deal with the term 'person' as though the possibility of an extratheoretic counterpart to it were as difficult a matter as in the case of 'black hole'. Hence the illusion that it is (in the terminology of the present book) somehow easier to be rationally aware of the place of the term 'person' in philosophical discourse and books of philosophy than it is to become rationally aware of a person. Philosophers off duty do not feel this problem, but an astonishing number insist that there is absolutely no way they can as philosophers be rationally aware of things as problematic as persons. Far easier to attend to such things as person slices; easier still to attend to a *language* that contains such expressions as 'person' and 'person slice'.

All this assumes that both poles of cyclical knowledge are focused on by direct rational awareness. Scientists who attend to and understand a body of theory as such already possess a knowledge different from the knowledge they would be said to possess about nature if that body of theory eventually passed appropriate empirical tests and came to be accepted. The body of theory itself is the focus of this different knowledge—theory as such is what is known, and known directly. And this means not only that it is understood all the while the appropriate symbolic structures are used articulately but also that it is then *experienced,* and experienced as a body of theory. To experience the theory as such is of course to attend to it against the background of commonsense experience, which no doubt enters into the texture of the theory in various subtle ways—ways more subtle than those by which theory is alleged to qualify experience. Odd as it may seem to say so, those who understand the theory are rationally *aware* of the theory as they understand it. Because direct rational awareness is our most essential way of knowing, problems about the tension between the rational and the empirical poles of science cannot legitimately be used to propound a fundamental epistemological problem about the accessibility of reality. The most fundamental epistemological problem about the relation between the rational and empirical poles of science cannot be addressed at the level of science, since an analogous but more basic relation exists between the rational and experiential poles of our direct rational awareness of a body of theory and of our direct rational awareness of whatever is relevant to the confirmation of the theory. The tension between the rational and *experiential* poles *within* rational awareness is more basic than the tension *between* the rational and *empirical* poles of cyclical knowing.

THESIS IV. In the case of science, despite contemporary claims to the contrary, whatever experience (empirical pole) tends to confirm or disconfirm a body of theory or part of it (rational pole) is not a function of that, or indeed any, body of theory.[3] The empirical pole of science can thus in principle be distinguished from the rational pole, as indeed it must be if the theoretic-empirical cycle is to be successful. The empirical pole is in fact what is usually called the commonsense world. Things, processes, and features of that world belong in principle to an independent level of reality (ontic level): they are not appearances constituted (in Kant's sense) either by our use of a certain language or by our holding, in some way heretofore unnoticed, a body of primitive and prescientific theory. Thus, when science, as it must, brings theory to the bar of commonsense experience, it is coping with something that is experience only from the point of view of a cyclical way of knowing. But, on the basis of our exercise in first philosophy, any item belonging to the empirical pole of science belongs to an ontic level that is attended to by (direct) rational awareness, a level directly known, and known by scientist and nonscientist alike.

From its beginnings down through the gradual perfection of the Newtonian system in the course of the eighteenth century, the growth of electromagnetic theory in the nineteenth century, and the revolutionary developments of our own century, modern science has been faithful to the

---

3. In Chapters 3 and 4, in which we were examining the linguistic consensus, I used the expression 'body of language-cum-theory'; but because we are now discussing radical realism, which rejects the conflation of language and theory, it is appropriate to use merely 'body of theory'.

theoretic-empirical cycle. It did not repudiate that cycle in the face of the internal criticism of empiricism we owe to Berkeley and Hume; nor did it alter its ways when that criticism awakened Kant from his dogmatic slumber, the slumber of his trust in the de facto empiricism of Newtonian science. And now, after the exponential increase in the power of both the "rational" and the "empirical" components in this theoretic-empirical cycle, and after all the twentieth-century controversy about the relation between those components, it still goes on in the same spirit. It does so, indeed, even for those scientists who, moved by the ambiguities of quantum theory, by the dominant antirealism of the linguistic consensus, or by both, devise instrumentation and experiments designed to bring out the reality-making power of observation guided by theory.

Nor has the estimate of the depth and penetration of science on the part of scientists, the many able publicists for science, or the intellectual community at large been much affected by what is said within the linguistic consensus about the relations between the members of the epistemic triad we considered in Chapter 4. Whatever experience may really be according to the consensus, whatever such a philosophy may tell us about whether an ideal body of scientific theory can provide us with knowledge of a reality lying behind, beneath, or above commonsense experience, the conviction is powerful within the scientific community that a body of scientific theory whose predictions are borne out within the world of commonsense experience can indeed be trusted to tell us how things extratheoretically are. With this goes the conviction that the world of common sense stands before science as an explicandum; that its usefulness as the locus of empirical testing is bound up with its status as explicandum; that for these same reasons the findings of common sense, just as they are in fact experienced, cannot be the whole truth about the way things extratheoretically are.

This faith has been there from the beginning of the scientific era. Today, despite the influence of antirealism among philosophers of science who belong to the linguistic consensus, it is fair to say that this faith has never been stronger within the working scientific community. For many scientists, it amounts to faith that an ideal body of scientific theory can tell us *everything* about how things really are. Hence the proliferation in recent years of bodies of theory that purport to be so comprehensive as to bring to a definitive end the whole business of theory making and theory testing—the whole business of the theoretic-empirical cycle. As noted earlier, there are many competing grand unified theories, but the spirit of optimism is so pervasive that it is widely assumed that, if none of these bodies of theory is adequate to the job, sooner or later some successor will be adequate to it. Hubristic or merely daring, such theories are realist in their telos: they are put forward as the ultimate truth about extratheoretic reality, not just as languages that enable us to cope pragmatically with the texture of commonsense experience.

There is therefore an important sense in which the achievements of science remain quite untouched by the concerns of philosophy of science. The theoretic-empirical cycle that begins in the commonsense world, moves away from it into a complex maze of theory building that seems to set common sense at naught, and finally returns to common sense for vindication or dismissal stands firm in the sense that, whatever philosophical interpretation we make of it, we shall not alter its essential nature. It is true that scientists themselves have sometimes felt that they must become epistemologists or even metaphysicians to provide satisfactory interpretations of what goes on when measurements or other instrumental interventions are made at the level of the exceedingly small and evanescent, but all this may be fairly said to mark a crisis in theory rather than a fundamental change in the mode of experience to which recourse is ultimately had. Whatever else theory may be, it is, as Stephen Hawking said, a thing of laws and equations and as such needs to be brought back to commonsense experience so that its worth may be known. And whatever commonsense experience is understood to be, consultation of it must go on in its own terms. For the working scientist, the theoretic-empirical cycle remains firm.

The difference between the two poles of the cycle also stands firm. Scientists cannot profit from the consultation of the empirical pole if they dismiss commonsense experience as either (a) a mode of *theorizing* about so-called stimuli that are not available *as* stimuli or (b) an experience that derives its very form from part of the same body of theory that needs to be put to an empirical test. And whatever scientists may think when they philosophize, they do not so dismiss it when they practice science. They dismiss or otherwise bracket the dismissal and then carry out the consultation. It is the body of scientific theory—that thing of laws and equations—that they wish to satisfy or adjust, not the epistemological theories of philosophers.

It is not difficult to distinguish a commonsense item as known by direct rational awareness from that same item interpreted in terms of some theory that is also an object of direct rational attention. It is true that if one interprets some commonsense article of experience in the light of theory (this mark on the photographic plate consists of molecules of a silver compound that has been exposed to light), it is then, qua so interpreted, theory-laden. More important, the article of experience becomes theory-laden whenever we regard it from the point of view of the theory for which it is alleged to be evidence (this mark, being just here rather than just a thought to the side, indicates a red shift of a spectral line, and this is consistent with the result required by a certain body of cosmological theory). But before either interpretation, a spot or a smudge is merely a commonsense article of experience, and it must be taken in as such before the interpretation in terms of theory can even begin. Furthermore, where it is invoked, as in the second case, because it bears on the adequacy of the theory, the invocation is relevant only to the degree that the very same commonsense experience could have been had in some other circumstance in which that body of theory, or indeed any body of theory, was not under consideration. In the case of the mark interpreted as a shifted spectral line, it is thus essential that someone indifferent to cosmological theory and indeed quite ignorant of it would, if asked, agree that the mark was just here rather than just there.

It is usual for philosophers who work within the linguistic consensus to agree that the very having of such an experience is cognitive, but only because it is in principle linguistic and so theoretic. Undergoing a commonsense experience is thus understood either as the interpretation of unknown stimuli by virtue of a theoretic-linguistic net of a primitive kind or as the outcome of a formative linguistic capacity thought to operate in a way for which Kant is offhandedly invoked as a precedent. On this view, the use of words like 'mark' and 'smudge'—and indeed of any substantive that does service in common language—amounts to the postulation of an entity as a way of making sense of stimuli we are bombarded by constantly even though we cannot experience the stimuli as such. But this approach can scarcely be used in support of the claim that all experience is theory-laden, since it begs the question that is at issue.[4]

Scientists who bring a body of theory to an empirical test must eventually do so by way of some commonsense experience, no matter what elaborate chain of reasoning convinces them that experience does indeed test that body of theory. The linguistic-consensus claim that the whole of the commonsense world is a function of a body of language-cum-theory is replaced in radical realism by the claim that each item of the commonsense world, as well as that world itself, is in principle something known by virtue of direct rational awareness. The consultation of commonsense experience is in fact a cognitive consultation: commonsense experience is not experience without qualification; it is not

---

4. Notice that the doctrine that commonsense objects belong to the ontology of commonsense language in the sense that they are postulates of the language is itself a theory about the commonsense world; further, these very stimuli belong to the ontology of the theory.

experience in an unqualified sense; it is commonsense *knowledge,* but here taken as the empirical pole of the theoretic-empirical cycle. To say so is not to pretend to settle the relative value of scientific and commonsense knowledge, but it does help to make it clear that scientific theory and the relevant experience are cognitively distinguishable—at least as distinguishable as figure and ground. It also suggests that, far from commonsense experience being a function of theory, there is a profound sense in which theory is a function of commonsense reality. This does not, however, alter the plain fact that acceptance of scientific theory often affects our assessment of the value and significance of commonsense knowledge.

THESIS V. A scientific body of theory includes (among other things) propositional structures that are formed (produced, made, created, constituted) by the formative function of rationality. This formative power of rationality by no means creates ex nihilo but rather on the basis of (a) (direct) rational awareness of that which is independently real, including among other things whatever belongs to the commonsense world; (b) whatever inner attunement with the real in general rationality possesses by virtue of being itself an expression of the U-factor, or common unity of Being—hence the permanent usefulness of the innateness principle, despite the many overstatements made in its favor; and (c) direct rational awareness of earlier bodies of theory. Features (a) and (b) help explain at least two puzzling features of all well-constructed theories, both those that turn out to be empirically adequate and those that do not: they always include some knowledge of the real even if they fail some vital empirical test (logic and mathematics, even of the most formal sort, give us *some* knowledge of the real), and they are often highly persuasive long before empirical confirmation or disconfirmation is available. A well-constructed body of theory includes many items that are, whatever else they may be, well-founded *entia rationis.*

A scientific body of theory consists of an intricate variety of interrelated structures that are at least in part dependent on the rationality of scientists. An unsystematic and incomplete catalogue of such structures would include doctrines, principles, theories (in a more restricted sense than in the expression 'body of theory'), hypotheses, postulates, formal systems, models, mathematical objects, and (formulations of) laws. Because they are all expressed with the help of propositions, these structures are in some sense also linguistic items, whatever other ontological status they may have. Indeed, contemporary epistemological fashion sees the totality of such structures as a linguistic network. But there are others who feel that an account of these structures in terms of concepts, ideas, or even essences is more fundamental than the linguistic one. For the purpose of this set of theses, all these structures are *entia rationis*. This term merely calls attention to the fact that the greater part of them are not available to be attended to without the rational acts of scientists. Our use of it leaves open the question what precise mix of defining, abstracting, constructing, creating, and discovering goes into these acts.

The production of any distinguishable *ens rationis*—for instance, the invention of some mathematical object in the course of the development of some new branch of mathematics—is guided by certain ideals, such as consistency, coherence, and completeness. The ideal of deductive unity for the whole body of theory constituted by all the *entia rationis* of science has been of great historical importance, and though it is now unfashionable among philosophers it is obviously still very much alive in physics. In any event, any well-developed structure of scientific *entia rationis* constituting a body of theory that both exhibits these ideals internally and coheres with a wider body of accepted theory makes strong truth claims on us even before it is put to empirical test. Nonetheless, this rational feature, or any part of it, remains incomplete, ungrounded, and unjustified without an appropriate relation to the empirical feature. The scientist's grasp of an empirically ungrounded theory is knowledge of a body of theory and—with the reservations expressed in Thesis V—only that. This claim is unaffected by the fact that the precise nature of the relation between a body of theory and the mode of experience relevant to it is a matter of philosophical debate.

It will be observed that, according to (a) of Thesis V, the empirical pole is relevant not only to the confirmation and disconfirmation of bodies of theory but also to their construction. It will also be clear that (a) and (b) together give well-constructed bodies of theory at least a minimal hold on the real. But that hold is not what is usually at issue in the realism-antirealism debate within philosophy of science. The next thesis sums up the findings of radical realism on that matter.

THESIS VI. Some empirically successful bodies of scientific theory require a realist interpretation (in a stronger sense than that of Thesis V) and others an antirealist interpretation. Empirical success is no doubt a matter of degree, but any successful body of theory will in the long run lead, however circuitously, to significant predictions at the empirical pole. If, in addition to whatever predictive value the body of theory has, it can also be shown that there is some isomorphism, however incomplete, between entities that belong to the body of theory and entities that are extratheoretic, then a realist interpretation is in order. If, however, there is predictive success but no isomorphisms of that kind can be shown, an antirealist interpretation is in order. Both realist and antirealist theories can save some of the phenomena. The arguments that purport to show that isomorphisms do or do not exist are complex, and in most cases the best we can hope for is indirect knowledge of what is in fact the case about such isomorphisms. Nonetheless, our knowledge of the body of theory as such is direct, and our arguments about whether that body of theory is to be understood realistically or antirealistically rest on direct knowing. So it is only by virtue of a radical realism established at the level of first philosophy that we can begin to decide whether a given body of theory should be given a realist or an antirealist interpretation.

The question whether a given body of theory possesses an isomorphism between theoretic and extratheoretic entities seems to be crucial to the senses of 'real' that are defended or dismissed in the realism-antirealism debate. For that reason, I have made it central to the formulation of Thesis VI. Nevertheless, there is another and more modest isomorphism between certain bodies of theory and the world of common sense that is worth considering. When predictions about the parameters of commonsense items belonging to the empirical pole of science are successful, those parameters are isomorphic with parameters deducible from a body of theory when it is applied to the empirical situation in question. Indeed, phenomena are saved only in the sense that an isomorphism of that kind has been established; and when they—or some of them—are saved, the body of theory in question has surely provided some indirect knowledge of measure properties of the commonsense world, whatever else it has done. If, as radical realism maintains, that world constitutes an ontic level that is extratheoretically real, the body of theory may therefore be said to be realistic in the limited sense that it gives some knowledge of the same level known to be real by virtue of direct knowing. No well-constructed theory of some empirical worth fails to give *some* report on the real, even if it has the antirealist failing of postulating the existence of entities that do not in fact exist—the failing, that is, that its ontology is different from the ontology of reality. According to Aristotelian astronomy, for instance, which is not without some empirical worth, there are indeed crystalline spheres, but in a reasonably straightforward sense of 'are' it appears that in fact there are none. To put the matter another way, although an empirically respectable body of theory, including its models, may not as a whole be isomorphic with what it purports to be about, it does include an isomorphism with some features of the real world. But this is simply another way of calling attention to the reality of the ontic level of the commonsense world. It is an important point, but not the point consensus realists wish to make.

THESIS VII. The plausibility of antirealism, and hence the very possibility of the realism-antirealism debate, depends on three of our capacities that are closely related. First, and most important, is our capacity for being rationally aware of two quite different kinds of things: independently real things that we have in no sense formed or constructed, and a great variety of other things that are in fact products of the formative function of rationality. Second is our ability to shift our attention rapidly from one object of attention to another without quite losing peripheral awareness of the others. Third is our power to deceive ourselves about what we are doing. All the entities that are products of the formative function are real in the sense that they are what they have been formed to be; they are real in the additional sense that rationality must draw on the independently real to form them; and, sometimes, they are also real in the sense that there are significant isomorphisms between them and the independently real. It is a commonplace failing, a failing we do not have to be philosophers to commit, to take such theories and imaginative constructions to be realities that are independent of our formative powers. When we are in this condition of error, we take certain entities to be real without qualification which, judged by the criterion of independence of our formative power, are merely Pickwickian realities. Antirealists insist that this condition is not what most of us take it to be—a condition of error, illusion, or self-deception—but rather our normal epistemic condition. For them, our-fundamental mistake is to suppose that there are (independent) realities over and above the "realities" we are fated to form by the very act of attending.

The most important source of error for human beings is their capacity to be rationally aware of, to know directly, both realities that are quite independent of their rationality and other realities whose reality is that of something formed by rationality. The latter are real qua constructions: in some important respects they are *entia rationis*, a category here understood to include concepts, merely linguistic items, bodies of theory, and imaginative constructions of many kinds. We are thus at the mercy of our own creativity: we suffer from an embarrassment of riches in our mental lives, and we take some of these constructed things to be what they are not. Our temptation to do this is strongest where there are things we wish to know that cannot be attended to directly: instrumentalities designed to provide indirect knowledge are taken either to *be* the things we want to know or else to provide us with an adequate representation of them.

Although one might expect science to be especially vulnerable to this temptation, it seems that the progress of science over the long term has not been much affected by it. It is true that there have been many scientific theories whose only ultimate worth is that they are interesting and plausible *entia rationis*. Nonetheless, science has been spared, in the long run, the most serious consequences of supposing a striking theory to provide a true account of (to provide indirect knowledge of) an independent reality when in fact it does not; for sooner or later scientists must bring their bodies of theory to confrontation with an empirical pole that is (in the sense of Thesis IV) independent of those theories.

But the disciplines that are cognitive in aspiration but not empirical in the way science is have not been spared; for there are many things besides the microentities of the physicist—important aspects of human motivation, for instance—that are hidden from our direct knowledge, and our ways of determining the adequacy of theory to such matters are often unsatisfactory. Nor are any of us in our private occasions spared the consequences of our prolific production of *entia rationis,* understanding that expression now in a sense broad enough to include a variety of imaginative constructions. More often than we like to think, we prefer the products of our teeming fantasy to reality, which is often hard to understand and accept as it is. Most of the time that preference does not have the justification it sometimes has in the case of the arts.

Philosophy belongs with these other disciplines, and its vulnerability is nowhere so clear as in the case of the doctrine of antirealism, which, in its extreme forms at any rate, tries to make a virtue out of our multiform power of self-deception.[5] It does so by proposing a theory about the human epistemic condition in which that power of self-deception is *defined* as the human condition—defined in the sense that knowing becomes, within that body of theory, worldmaking or reality making. Philosophy of that kind thus provides in advance a justification for a tide of relativism generated by forces of which that philosophy is probably more symptom than cause. The more extreme forms of this relativism have provided many a lucrative career both outside and inside academic philosophy for those whose unfailing reality instinct for the main chance provides them with one canny insight into the human condition: that, by many people for whom culture is a recreation, a paradox whose sources are not understood is taken to be a profound truth. Thus one careerist tells us that, seeing that truth is a myth, the ultimate obligation of the investigator is not to say something true but something interesting; another tells us that the so-called pursuit of truth is in fact the pursuit of power—not just sometimes, as it surely is, but always. And both careerists, well aware that when they say so they are claiming to say what is the case about the human

---

5. Remember that linguistic consensus realists are realists only about science. See Chapter 4 as well as the preliminary discussion of this matter in Chapter 1, Section 3.

condition—and what is the case quite apart from their making it so by a cultural power play of their own—exemplify this paradoxical oscillation to the astonished but trusting cultural consumer.

THESIS VIII. It is rational agents who exercise the function of (direct) rational awareness and who make or produce not only bodies of theory but explanations in general. As agents, they—who are also our particular selves—are well within the range of our direct knowing, even though their infrastructures—their physiology, for instance—and the internal aspects of their minds are not. The ontic level of the rational agent, that is to say, is not only a fit subject for philosophical and scientific investigations; it is also the source of all investigation and of all other creative activity. Various reductive challenges to the ontological status of rational agents are thus in principle dubious. Conversely, the "circle" in discussions like the present one—rational awareness being used in support of the authenticity of agency while agency is deployed in support of the authenticity of rational awareness—is a benign one.

Although this thesis sums up one of the most important themes of this book, it is a theme I have explored extensively in earlier books and shall be returning to in the next chapter. It probably serves its purpose well enough without further elaboration.

THESIS IX. Direct knowledge, authenticated and deepened by its own naturally reflexive component, is potentially first philosophy. Its hold on reality, or being, is one of active discovery and achievement rather than passive reception.[6] One by-product of this kind of reflection is

---

6. This theme is extensively developed in Edward Pols, *The Recognition of Reason* (Carbondale: Southern Illinois University Press, 1963). The theme

the vindication of much that belongs to commonsense direct knowledge: it now appears as the settled outcome of past direct cognitive achievements. Speculative systems that are in effect metaphysical bodies of theory can be misleading if they fail to accord with the findings of reflexively intensified direct knowledge.

Metaphysics has always consisted in great part of theories, and it is well known that in some eras these theories were also intended to do the work now done by the bodies of theory of the physical sciences. Philosophers have been aware for some time that the exponential pace of scientific development has left metaphysical theory makers with less and less to do. Perhaps it was this awareness, coupled with the realization that metaphysical theories cannot be brought to an empirical pole in the same decisive way as can most scientific theories, that encouraged the fashion of speaking of metaphysical theories as speculative. Whitehead, who speaks of his own metaphysical work as speculative philosophy, is probably the most influential figure in this development. In any event, the expressions 'systematic metaphysics' and 'speculative metaphysics' are now generally interchangeable in those provinces of the academy in which metaphysics is still regarded as respectable. It seems to me that the most distinctive contribution that can be made by metaphysics is at once more modest and more momentous than the construction of speculative systems; and that is another reason—besides those given in the second chapter—for my preferring Aristotle's original name for the discipline, first philosophy. My views on the positive nature of what first philosophy can do are by now probably clear enough; in any event, they will be by the end of the next chapter, which is chiefly concerned with first philosophy. Just here I am more concerned with the danger of forming a speculative body of theory that is clearly a mere linguistic-

---

of recognition in that book is closely related to the theme of direct knowledge, but the earlier book is less precise about the second basic function of our rationality (the formative one) than I trust this present book to be.

cum-theoretic reality and preferring it to the reality, including the reality of ourselves as rational agents, that we are in a position to know directly, although by no means exhaustively. Speculative theories that respect this reality and even use it as a point of departure are another matter, and if we manage to assert the control of a responsible first philosophy over speculation, there may be a future for such enterprises.

7

- First Philosophy and the Reflexivity of Direct Knowing

1 *"The Science We Are Seeking":*
   Bringing First Philosophy into Being

- First philosophy is always under an obligation to justify itself: its substantive topics can be legitimately pursued only if it is also shown in that very pursuit by what authority, by what right, the philosopher does so. Indeed, the first task of first philosophy—first in order of importance, even if it happens to be performed simultaneously with other tasks—is self-justification. Although the topics actually discussed by philosophers who are doing first philosophy (in somewhat the sense we examined in Chapter 2) suggest that they feel an obligation of that kind, the nature of the obligation has never been satisfactorily articulated. It is therefore not surprising that a satisfactory justification has never been provided.

If, proceeding on the assumption that there is an obligation of that kind, we should ever manage to respond to it adequately and thus achieve a first philosophy that commanded general respect, it would then be seen to be first not just in its substantive findings but in its rational authority as well. That prospect is enough to make me wish to persuade all those who are interested in bringing the discipline into actuality to call it, as I do here, by its earliest and most illuminating name. And it is no mere rhetorical flourish to speak of actuality, for the discipline has never achieved the pervasive and living authority that continues to be our deepest intellectual need; it is still what Aristotle called it—"the science we are seeking." There have been any number of metaphysical systems—in the terminology of this book, metaphysical bodies of theory—and they have been filled with powerful suggestions about the structure of all the things and all the features of reality we are in no position to know directly. But these systems have given us little reassurance about our capacity to know anything at all as it is—things that are well within our range no less than things that are hidden—and so we are inclined to distrust them.

In language appropriate to our present investigation, we may say that those who devoted themselves to the discipline of first philosophy should have systematically cultivated our capacity for direct rational awareness of being, or reality, as it is in its independence of our wish or will on the one hand and of whatever formative capacities our reason may possess on the other; and that they should have relied on the reflexive intensification of that capacity to provide its own authentication. It has been the thesis of this book that they could have done so only if, in the course of philosophic inquiry, they had recognized that in all their daily occasions they were already exercising a capacity to know directly things that are independently real in the sense just mentioned. Long before philosophy raised the reality question, these philosophers, and the rest of humanity as well, were enjoying a commonplace epiphany—a rational-experiential satisfaction in their own attainment of something that was independently real. But they had not yet adequately realized what they were doing; and to the degree the matter remained there, with no reflexive development of the achievement, the epiphany remained unrecognized and unacknowledged.

A reflexive turn in which direct rational awareness attends to its own rational-experiential engagement with the independently real is therefore essential to get first philosophy under way. But it then takes some persistence to recognize that in having a direct cognitive hold on some few real things we already have a bridgehead in the real in general. As I said in Chapter 2, long before the Latin-based term 'real' became current in philosophy, Greek philosophers were using various inflectional forms of the verb 'to be' in dealing with questions related to the subject matter of this book. So let me repeat what I have just said, this time using that older word: it takes some persistence to recognize that in having a direct cognitive hold on some few beings we already have a bridgehead in Being. Philosophers who take first philosophy seriously have never abandoned the vocabulary of being, so this way of expressing the matter is an effective reminder that today's realism-antirealism controversy is in fact a controversy in first philosophy.

There is another and more subtle reason for speaking here of beings and Being. The term 'beings' emphasizes particularity, the term 'Being' generality; yet the two terms are in fact one, physically distinguished only by the sign of the plural on the first and a capital letter on the second, and otherwise distinguished only by their different emphases. In the background is another term, 'U-factor', to remind us that each particular thing is the concrete bearer of universality/unity and that Being, together with the U-factor, which partially characterizes it, is integral to the particulars it transcends and so is not present to our rational awareness as an abstraction. It is not, of course, the *terms* that primarily concern us here but rather, as radical realism requires, what the terms call our attention to.

When we attend to realities and the Real—to beings and Being—it becomes clear that when we say that 'real' means 'ontological independence of any formative power exercised by the knower' we do not say all there is to say about the reality question. Being, we now see, admits of degree. Although all realities/beings are alike in being independent of something formative in any act of rational awareness that attains them, they are not alike in comprehensiveness, power, causal originality, and completeness. On the one hand, the person who exercises rational awareness to know a stone or tree directly is no more real than the stone or the tree. The person has an independent reality that, qua known, is not formed by the rational awareness of a second person who attends to it; but that is equally true of the stone and the tree; measured by the standard of ontological independence of a knower, those realities—person, stone, and tree—form a democracy. On the other hand, persons exercise a cognitive power and causal originality to know things like stones and trees, and this seems to warrant our assigning them to a higher ontic level—a higher level of Being—than those of the stone or the tree. One sign of this is the great difference between the presence of the U-factor in such things as stones and trees and its presence in the person who is rationally aware of stone or tree. It is the difference between merely exemplifying the U-factor and exemplifying the U-factor while also being rationally aware of the U-factor and of the unique affinity between rational awareness and the U-factor. The U-factor, moreover, is the factor in Being that warrants our talking of degrees of Being. When we arrange a number of beings in order of their ontological rank it is the presence of the U-factor in each of them and in Being as well that makes them participants in Being; they are rightly ordered as particulars because of the degree to which they participate in what is not a particular.

In any event, once a bridgehead is secured in Being, we feel obliged to intensify our reflexive exercise of rational awareness and thus enlarge the bridgehead. The obligation of first philosophy to justify itself is in the long run an obligation to know Being directly as well as we can. How far we shall be able to go we do not know; an assured rational awareness of some independent beings and the beginning of a rational awareness of what they participate in give us no assurance that we shall eventually be able to scan Being to all its height and depth. But the possibility of a gradual growth in our power of direct knowing is there the moment the reality of the function of rational awareness dawns upon us in the course of our reflexive development of it.

That is the reflexive epiphany mentioned in Chapter 2, and it brings us a rational-experiential satisfaction in the U-factor of Being, integral with and sustaining the particularity of each of the independently real beings attained in the commonplace epiphany—sustaining, indeed, each moment in the development of the function of rational awareness as well, for rational awareness is a real function of a real being. Cultivating this reflexive epiphany, we produce no foundationally final propositions, but we do use propositions that emerge from this matrix to guide us back to the achievements of rational awareness itself. Its first achievement is prior to philosophical reflection, for the commonplace epiphany is truly commonplace: when we are engaged in rational awareness of this kind, we are so taken up with the independent nature of whatever we are attending to that it does not as yet occur to us that knowing some particular being directly, in all its independence of our cognitive powers, is itself a creative achievement. Our first concern in reflexive intensification is to remain faithful to the commonplace epiphany, holding fast to whatever reality it has achieved and then bringing into reflexive view the reality of the achieving function as well. The deployment and redeployment of the reflexive power of rational awareness to attain, on the one hand, real things and the Reality/Being they participate in and, on the other, the real function that accomplishes this feat yields the only warrant we can have and the only one we need.

One profound difficulty about this reflexive exercise is that reflexivity itself is part of its subject matter. As we try to be faithful to that important aspect of our rational awareness, various metaphors—creatures of the formative function of rationality—insinuate themselves between ourselves and the great complexity of what we are trying to attend to. I have myself used the metaphor of a cognitive knot and the metaphor of a cognitive circle to deal with certain aspects of our reflexive deployment of rational awareness. But the most common metaphor for a somewhat different aspect of it is the one in which the person reflecting is assumed to be enclosed in consciousness and to be seeing there, as in a mirror, ideas or images that purport to represent real things; it may be the most misleading of all the metaphors.[1] But the intimacy, finality, and unity of the reflexiveness of our rational awareness seem to evade us whenever we provide a metaphorical substitute for it, no matter how ingenious our invention. Evidently our first task in first philosophy is to learn to trust the reflexive authority of the function of rational awareness and not to allow the formative function of rationality to interpose its metaphorical products, not to speak of its model-like and theoretic products, as false images of that authority.

Although no Cartesian, I am moved to think here of Descartes—to think not of the disembodied clarity that was his explicit ideal but rather of a different rational authority he sometimes seems to have been reaching for. Powerful received doctrines hindered him: the representative theory of ideas, the cognitive primacy of propositions, the revived skepticism of the ancients, the ideal of knowledge as a deductive unity. Distracted by these as he was, he seems nonetheless to have been groping for the authority-in-act of a rational being existing in this world, but no less authoritative for that—a being whose authoritative reason is on the one hand an instance of Being and on the other capable of knowing Being as it is. It is an authority which—if real—will never be completely captured in a proposition, not even a proposition as "existential" as the cogito purports to be.

But if that is the first task of first philosophy, it is by no means the last. Some of the further tasks are substantive in

---

1. This form of the mirror metaphor (as applied to knowledge in general) becomes important only after Descartes's statement of the doctrine of representative ideas. Thereafter it occurs just as often in an analytic-empirical setting as in a metaphysical one. I am far from agreeing with Richard Rorty's account of the matter in his *Philosophy and the Mirror of Nature* (Princeton: Princeton University Press, 1979); see Edward Pols, *The Acts of Our Being: A Reflection on Agency and Responsibility* (Amherst: University of Massachusetts Press), pp. 220–25, nn. 2–3.

that they do not concern our cognitive authority except insofar as our need to accomplish them incites us to seek the authority to do so or to set them aside as not feasible.[2] In what follows I will press on with two of them. The tasks I have in mind have classic precedents in the problems of causality and substance, although the latter problem is surely misnamed, as I argue toward the end of the next section. But because the name was used by the great rationalists to discuss what they claimed was a knowable reality, by the empiricists in their rejection of that claim, and by Kant in his attempt to find a middle ground, it seems best to use it here and so give the problem a familiar setting at its first appearance. Radical realism, however, suggests that we should use the term 'primary being' (or 'primary entity') instead of 'substance'. Beginning with the title of the next section, I therefore use those new terms, reverting to the term 'substance' only when I have some historical precedent in mind.

2 Two Substantive Tasks of First Philosophy: Knowing Causality and Primary Beings Directly

Nothing is more scandalous about radical realism than its attitude toward the classic problems of our knowledge of causality and substance. Although it concedes that these things are not easy to know in all their nuances, it also claims that there is no problem at all about our capacity to know them directly and in a progressively more penetrating way. There are any number of problems about whether we can know certain principles to be true that purport to sum up the causal relation or to sum up what it is to be a substance. Radical realism claims that some of these principles—they are, of

[2]. I am aware of the oddness of using the term 'substantive' about these tasks when I am about to suggest that the term 'substance' is a misleading name for what one of the tasks is concerned with. But the term 'substantive' is well established outside the field of philosophy, and it seems just right for distinguishing the status of these tasks from more methodological ones.

course, propositions—can be derived, with more or less adequacy, from our direct rational awareness of real beings in interaction, interdependence, and interpenetration. But no principles of that kind are true without qualification, for whatever truth they do have goes back to the adequacy of our direct knowing of those real beings and the real relations between them. And naturally any qualifications we are able to provide must also emerge from direct rational awareness.

Consider one great causal principle: the principle of sufficient reason as expressed by Leibniz. Few philosophers take the primacy of propositions more for granted than Leibniz does, for he holds that the very foundation of all knowledge consists in principles known to be true.[3]

> As a man who wishes to construct a building on solid ground must continue to dig with his spade until he comes to a solid and stony basis, and as Archimedes required an immovable point in order to be able to lift the universe;—so we are in need of a fixed point as a foundation upon which we may establish the elements of human knowledge. And this starting point is the analysis of the different kinds of truth.[4]

For Leibniz, one such fundamental truth is the principle of sufficient reason; indeed, he claims that a greater reliance on this principle is essential to the sound development of *prima philosophia*.[5] It is significant, however, that the allegedly foundational character of this principle is intimately linked in his

---

3. In his time the power of the foundation metaphor was felt by most major writers, so I must take account of it here, despite all the reservations I have already expressed about it.

4. Gottfried Leibniz, *Opuscules et fragments inédits*, ed. Louis Couturat (Paris, 1903), p. 401; as quoted in Ernst Cassirer, "Newton and Leibniz," *Philosophical Review*, 52 (1943): 366–91, p. 374.

5. "It must be confessed, that though this great principle (of sufficient reason) has been acknowledged, yet it has not been sufficiently made use of. Which is, in great measure, the reason why the *prima philosophia* has not been hitherto so fruitful and demonstrative, as it should have been." *The Leibniz-Clarke Correspondence*, ed. H. G. Alexander (Manchester: Manchester University Press, 1956), pp. 60–61.

mind with the principle of contradiction. I quote the two principles as they appear together in the *Monadology:*

> 31. Our reasonings are founded on *two great principles, that of contradiction,* in virtue of which we judge that to be *false* which involves contradiction, and that *true,* which is opposed or contradictory to the false.
>
> 32. And *that of sufficient reason,* in virtue of which we hold that no fact can be real or existent, no statement true, unless there be a sufficient reason why it is so and not otherwise, although most often these reasons cannot be known to us.[6]

The principle of sufficient reason, Leibniz thinks, has its chief application in the realm of experience: knowing the principle to be true, we must look to the contingent facts of some particular empirical situation to determine the cause or causes of an event—that is, the reasons that explain it. But the principle itself, he thinks, is no more derivable from experience than is the principle of contradiction. Indeed, he claims that we know the principle of sufficient reason to be true on the basis of the principle of contradiction, although he insists that this dependence in no way alters the contingency of the causal connections we look for with the help of the former principle. "I certainly maintain," he says in a letter to des Bosses, "that a power of determining oneself without any cause or without any source of determination, implies contradiction, as does a relation without foundations; but from this the metaphysical necessity of all effects does not follow. For it suffices that the cause or reason be not one that metaphysically necessitates, though it is metaphysically necessary that there should be some such cause."[7]

For a full account of the principle of sufficient reason as Leibniz understands it, we must go to the rest of his system—

---

6. As translated in *Leibniz: The Monadology and Other Philosophical Writings,* ed. R. Latta (Oxford: Oxford University Press, 1898).

7. Leibniz to des Bosses, *Die philosophischen Schriften von G. W. Leibniz,* ed. C. J. Gerhardt (Berlin, 1875–90), vol. 2, p. 420, as quoted by Bertrand Russell, *A Critical Exposition of the Philosophy of Leibniz* (Cambridge: Cambridge University Press, 1900), p. 35.

that is, the rest of the complex of propositions that make up his body of philosophical theory. The main bearings of the principle within the system are familiar enough. Monads, which he also calls substances, can in principle be analyzed in terms of predicates attaching to a subject: they are, in fact, the subjects to which the predicates belong, and a complete knowledge of one of them would thus yield all of its predicates. But since all monads are also eternal, a complete knowledge of one of them from that perspective would be both analytic and a priori. An analysis of that kind would, however, be infinite and so could be carried out only by a divine mind. 'Analytic' must thus be taken in an unusual sense in this context, for the attachment of predicate to subject would still (in accordance with the principle) be contingent rather than necessary, seeing that, as Leibniz says, there is no contradiction in imagining the opposite of some particular predicate. Leibniz supposes, moreover, that the ordering of the whole causal chain depends on God's understanding of what is best or most fitting; so the nature of each particular contingent causal connection must be understood in the same way. In sum, what is true on the principle of sufficient reason is, from God's point of view, contingent, a priori, and analytic (but not finitely); from our point of view, however, it is contingent, a posteriori, and (as it is usual to say since Kant) synthetic. In both cases contingency is tied to goodness.

Although in the following discussion of causality I propose to use Leibniz's statement of the principle of sufficient reason to make a certain point about the appropriate source of causal principles in general, it should be remembered that there are many other formulations of this principle. Plato is responsible for one of the earliest, and his version of it—setting aside the systematic connections of the statement with the rest of his own philosophy—has a good deal in common with Leibniz's. I quote the statement of it in the *Timaeus*, followed by a similar one from the *Philebus*. The latter will be clearer if we remember that it is in effect a question, and that it is immediately answered with a yes.

Each thing that becomes becomes of necessity by virtue of a cause; for without a cause all coming to be is impossible.[8]

See whether it seems to you to be necessary that all things that come into being do so through a cause.[9]

In Plato's version there is the same emphasis as in Leibniz's on the necessity that there be a cause; and in the Greek, more clearly than in our word 'cause', a cause is understood to be a reason. As to the connection of the notion of a cause with the rest of Plato's doctrine, the primary positive sense of the notion always involves the Good; in his later work—especially the *Timaeus* and the *Philebus*—the Maker looks toward the Good as the goal of his making. There are, to be sure, important differences from Leibniz: the Maker is not the Creator of Leibniz's Christianity; and Plato's conditioning (or accessory) causes are said to operate with necessity—a different sort of necessity, presumably, from the rational necessity that makes us able to say, as in the passages quoted, that it is necessary that some cause operate in coming to be; and so the Maker must try to persuade them to his ends.

But I wish to take Leibniz's statement of the principle as my primary example, and I intend to detach it from the rest of his system to make a certain point. So we can make a similar detachment in the case of Plato and note that the principles so detached have much in common, although over two thousand years separate them. Turning, then, to Leibniz's principle, we exclude his subject-predicate account of causality, the predestined harmony, the goodness and perfection of God as the ultimate sufficient reason why things are as they

---

8. *Timaeus* 28A. The translation of this brief passage is my own. Cornford's translation introduces the notion of agency, an appropriate enough interpretation, since the mythical setting is developed in terms of a Maker. But it *is* an interpretation: the notion of agency does not actually occur in the passage I quote. Bury also translates without the word 'agency'.

9. *Philebus* 26E. The translation is my own; once again Plato's setting, as well as words that immediately follow, require that a cause be in the long run a maker or agent; but that is not to our immediate purpose.

are, and certain other themes familiar from either the *Monadology* or the *Discourse on Metaphysics*. After this exclusion we are still left with a minimal principle whose credentials are powerful. Indeed, first philosophy aside, it would appear that some such minimal principle of sufficient reason sums up the heuristic conviction that guides most scientific investigation. No doubt within that setting it is used with certain qualifications. Modern cosmologists and other searchers for a grand unified theory, for instance, tend to take the "big bang" as the ultimate sufficient reason and to suppose that it is pointless to ask for a sufficient reason for *that*. On the other hand, if we go back to Leibniz's version, restoring for the moment all the other propositions with which his statement of the principle makes coherent connection, we remember that he too qualifies the principle's scope: God, understood as the Perfect Being, is the ultimate sufficient reason.

Our present concern, however, is with a "thin" version of the principle. Although, unlike Leibniz, I do not think we can find a final authority for *prima philosophia* in the universality and necessity of that principle or any other, I ask the reader to consider whether some such principle does not emerge out of our direct rational awareness of beings in their causal intercourse. If it does, the *principle* is not our authority, but rather what we are rationally aware of, together with the active power of rational awareness itself. As for the principle, though it might be adequate in some degree to what we are rationally aware of, it is always revisable by an increase in the exactness and intensity of rational awareness and so to that degree is inadequate to our rational awareness of the causal relation. Certainly no particular formulation of the principle is unqualifiedly universal and necessary. We must look elsewhere for universality and necessity. Rational awareness is always directed on real beings, which in respect of their common U-factor are participants in Being; and it is Being, rather than a principle partially and tentatively expressing some feature of Being, which is truly universal and necessary.

The "thin" principle, emerging thus from our rational awareness of real things, may be "thickened" in various ways. We can, for example, construe the principle in terms of a cause (or reason) that occurs before an effect or in terms of a cause (or reason) that pervades a whole temporospatial order with its efficacy and so may not properly be said to occur before an event within that order, an event which it nonetheless causes (or explains). Whether we have any right to construe causality in both ways depends on what rational awareness attains in attending to temporospatial beings in their relatedness with one another. But it should be noted that scientists routinely thicken the principle in the latter way as they use it heuristically: a continuous mathematical function that allows for the prediction of a state of a physical system on the basis of an earlier state is understood to explain the process by giving a reason why it develops just as it did. In this respect the thickening invokes a quasi-formal causality rather than the merely efficient kind invoked if one considers merely the state of a system at time$_1$ and its being succeeded by a state at time$_2$. No doubt telic doctrines like those of Plato and Leibniz thicken the principle in other ways. Superficially, it looks as though this happens because the notion 'cause' in their philosophical bodies of theory coheres with the notions 'purpose' and 'Maker' as they are expressed in various propositions of the systems. This way of putting it, however, suggesting as it does that those philosophers were merely speculative system builders, may do them less than justice.

In any event, radical realism suggests that there are more experiential ways of thickening the principle than by way of mere system building. And the reader, considering with me now a different approach to the status of causal principles, may even concede that an intensification of our rational awareness of things in causal concourse might someday yield a thickening of the principle in which the notion 'telos', freed from its excessively anthropomorphic resonance, would draw attention to something important that we now miss when the quasi-formal causality of scientific law completely dominates whatever in our current notion of causality concerns something that stands outside the sequence it explains.

Be that as it may, radical realism is far from relying on causal principles, thin or thick, that purport to carry their own warrants with them merely as propositions and to be independent of any experiential base. On the other hand, it is far from supposing that principles about causality and, for that matter, about what has traditionally been called substance cannot be formulated on the basis of our progressive intercourse with the real by virtue of rational awareness. For the status of such progressively modifiable principles we should probably have a new category. Certainly neither 'analytic a priori' nor 'synthetic a priori' will do and 'synthetic a posteriori' misses the radically innovative character as well as the tendency toward universality we might venture to ascribe to them on the basis of the foregoing discussion. I suggest 'rational-experiential', understood in terms of the several accounts of rational awareness I have provided in the course of this book.

The chief point of this consideration of the thin form of one great and persistent causal principle is that radical realism looks for its authority to (direct) rational awareness and to what rational awareness attains of the real in both the commonplace and the reflexive epiphanies—and that, although principles of various kinds may more or less sum up what we have won in relying on that authority, they are the outcome of the exercise of that authority and by no means the source of it. Keeping this point well in mind, we are ready to look at the two substantive tasks mentioned earlier. Although our consideration so far has been carried on with reference to a causal principle, we shall find that in practice the most important form of causality is intimately associated with the causal efficacy of temporospatial beings, which I am here calling primary beings. A consideration of such beings replaces the consideration of substance that has played so important a part in modern philosophy since Descartes, and those primary beings called human beings will figure prominently in this consideration, not, however, as the source of our notion of causality but rather as an instance of causality—perhaps the most important instance for first philosophy. The expression 'primary being' is, of course, a deliberate echo of Aristotle's *protē ousia;* indeed, I take it to be a better translation of that expression than is 'primary substance'.[10]

It is now widely believed that the self-identity of the human self (or the human person) cannot be that of a substance. When the term 'substance' is used in the sense current in modern philosophy, I share that belief. But some reservations are in order. For one thing, the rejected doctrine is expressed in terms of a word, 'substance', that carries over very little of the Greek word, *ousia,* it purports to translate; for another, the doctrine has been reworked considerably from the seventeenth century on. At least three important things have been lost: (a) the idea of being that is present in the Greek word by virtue of the fact that *ousia* is a substantive formed from a feminine participle of the verb 'to be'; (b) the dynamic overtones present in such Aristotelian expressions as *ousia energeia* and *ousia entelecheia;* and (c) Aristotle's insistence that a primary *ousia* is an individual being, a "this," rather than something so diffused and general as mind or matter—common instances of substance from the seventeenth century on. The full account of the human *ousia* in *De Anima* and the *Nichomachean Ethics,* in which the rationality of the agent shapes, by virtue of its choices and actions, its own coming-to-be, is a dynamic one. The self-integration of an entity of that kind is inseparable from the issue of its self-identity. Self-integration takes a considerable time, and its outcome is always much in doubt. Yet Aristotle calls this human self an *ousia*—a being, an entity—and I feel no hesitation in following him at least

10. In the past I have sometimes used the expressions 'actual being' and 'fundamental being', but I now think that 'primary being' is on the whole the best expression to use. Some of my reasons follow immediately in the text. On this terminological point, see also Pols, *Acts of Our Being,* pp. 192–98, 205–12, 229–30, nn. 7, 10.

that far. That is not to say that I consider the human person to be a substance in most of the senses the great philosophers from Descartes through Kant would have accepted. Leibniz's view of substance is, to be sure, more Aristotelian than that of most of those philosophers: the monad/substance is nothing if not a "this," and most of the time Leibniz attributes a dynamism to it. But there is another side to his philosophy—we glanced at it in considering his subject-predicate view of causality—in which the dynamism is in effect merely phenomenal. As to Aristotle, I do not want to bring in the whole of his doctrine of *ousia,* for there are undoubtedly elements in it that contributed to the post-Cartesian history of the concept of substance.[11]

3 The Unity of the Two Tasks: The Causality of Primary Beings

I ask the reader to recall the discussion in Chapter 5, Section 2, of what I called the primary mode of the function of rational awareness. I maintained there that, as we deploy the function of rational awareness toward temporospatial beings such as persons, animals, plants, and artifacts, we trace the temporospatial range of what we are attending to. We do this, moreover, from our own temporospatial situations. Our awareness itself takes up time and is suffused by the temporality of what it is engaged with; and from its own more or less adjacent spatial region it takes in the spatial region that is integral with the temporality of the thing attended to.

I mentioned also that for rational awareness to actualize itself in a direct knowing of the being in question it need not track or trace the whole temporospatial career of what it attends to. Seeing that the function is *rational* awareness, a small part of that range suffices for it to complete itself. It is rationality that actualizes itself *in* awareness, responding not just to the particularity of the item attended to but also to a factor in the item that transcends any part of the particularity

---

11. See ibid., pp. 192–94, 228–29, n. 3.

because it qualifies all such parts; that is what I have been calling the U-factor. Our tracing of a fragmentary part of the temporospatial totality of the thing thus gives all the experiential base we need to recognize the thing as participant in the U-factor common to all other temporospatial things, including those, like ourselves, who are also capable of rational awareness. Recognition thus embraces and assents to both that in the thing which is merely particular and that which transcends particularity, and it includes the recognition that the very "thisness" of the thing directly known depends upon its intimate association with what is not a "thisness." Our assurance that what we thus attend to does indeed have the independent reality status this book has been concerned with depends, as we have seen, on a reflexive turn—a reflexive epiphany—in which the importance of the U-factor in particularity becomes salient.

I want to make it clear in this section that the two concrete tasks we are now concerned with are in fact one: the appropriate source of what philosophers usually discuss under the rubric of causality is absolutely integral with the temporospatial beings we attend to in primary rational awareness, so integral, indeed, that to attend to a being as a being is also to attend to a causality operative throughout its temporospatial deployment. For the moment we are not considering the way a primary being operates on the world about it and so is a cause within that world. We are considering rather its own internal constitution and finding that constitution itself to be causal, and causal in a more original sense than that in which the being operates as a cause on other things in its world. The development and deployment of the being of a primary being is coincidental with the deployment of a power throughout its structure. It is a paradoxical power, for it is exercised on a multiplicity of beings (many of them primary beings) that are in one sense components of its being and in another sense distinct from its being. It is a power that *causes* those components to be components, for without that power they would pursue independent courses of their own. We are, then, considering the unity of the primary being to which we attribute this power—call it the superordinate primary being—and contrasting it with the vast multiplicity of subordinate beings that in one sense make it up and in another sense are *made up into it* by its power. In our own case, this is the contrast of the unity of our being as rational agents with such a multiplicity as that of our cells or that of the components of our cells. It is obviously a contrast that is not merely structural, for it has a temporal and developmental dimension that is of profound importance.

Because of the association of this most important sense of power with the very being of a primary being, I have in the past called it 'ontic power'. The point would have been clearer if I had also called it 'ontic causality', and so in this discussion I will use those expressions interchangeably. It is, however, not just *super*ordinate primary beings that deploy that ontic power or causality, for the *sub*ordinates under its sway are, in a qualified sense, components of the superordinate and so make an ontic contribution to the being of the latter by virtue of their own ontic power/causality. In earlier discussions I accordingly made a distinction between superordinate, or supervening, ontic power and conditioning ontic power. In those same discussions I contrasted ontic power, as constitutive of the very being of a primary being, with the being's exercise of power on the world about it, which I called 'causal power'. In the terminology I now propose, there is thus a contrast of 'ontic power/causality' with 'causal power', and this may be less than perspicuous. It may help if we draw on a useful medieval distinction that was applied more often to God than to finite entities: ontic power/causality is *immanent causality,* and causal power, being exercised on something else by virtue of immanent causality, is *transeunt causality.* I have hesitated to put the matter in just that way in the past because the term 'transeunt' has recently been current in a sense that misses the point of the medieval distinction.[12] Just here, however, what is important is that rational awareness, in attaining

12. I address that misunderstanding in ibid., pp. 31–34, 218–19, n. 3.

the independent reality of some other primary being, attains also the ontic power/causality immanent in that primary being. It follows that what we have been calling the U-factor has a causal aspect no less than aspects of universality and unity.

We move now to a further development of the reciprocal themes of being and causality and to our rational awareness of both as they are exemplified in rational agents like ourselves—and so exemplified also in the rational awareness we are relying on. It is sometimes supposed that if so-called agent causality is indeed authentic, it must be an anomalous intrusion into a natural order pervaded by a causality of another kind. That is not so, although only agents can exemplify causality in just the way I was concerned with in two earlier books in which the topics of moral choice and responsibility were central.[13] But the doctrine of causality set forth in those books, like the brief rehearsal of it here, was a truly general one. It applied not just to agents but to all of the class of beings here called primary beings; and it applied also, with appropriate adjustments, to mass processes involving myriads of primary beings that we are not capable of distinguishing directly as such. The doctrine is not a theory, if it is true that our rational awareness of causality is integral with our direct rational awareness of *some* primary beings; but it can nevertheless be given a theoretic, or speculative, extension to many things that can be known only indirectly.

General though the doctrine was, agents furnished the usual examples of causality in those books. I tried to attend directly, and to call on my readers to attend directly, to the unity of rational human action; and I insisted on the explanatory value such action has if our rational awareness of it in fact reveals something that is a true operant power and not a mere appearance. I applied the doctrine in some detail to the mind-body problem and to the ontic responsibility of agents,

---

13. See Edward Pols, *Meditation on a Prisoner: Towards Understanding Action and Mind* (Carbondale: Southern Illinois University Press, 1975), and *Acts of Our Being* as indexed under the terms 'causality', 'ontic power', 'causal power', 'cause', 'power', and so on.

which I take to be a precondition of moral responsibility. Since agents are clearly good examples of primary beings and their causality, it is appropriate enough to use them here as examples of how our rational awareness of primary beings marches with our rational awareness of their causality. It is also convenient, for it is rational beings who exercise rational awareness, and we must in due course apply our account of causality to rational awareness itself.

Let me take my own actions in writing what follows as an example. But since (direct) rational awareness—my own and the reader's—is our theme, let us once again imagine that we are in conversation together and that I have embarked on a further explanation of the doctrine of causality we have just been discussing.

As I speak these words, I exercise power in the world about me. My action is an exercise of physical power of which you and I are both rationally aware. Science tells me that I am producing sound waves, which, by way of your eardrums, excite your nervous system. If you understand my words, I also exercise a power of a somewhat different kind on your mind. But let us neglect that kind for a moment; what I am getting at just now is that by my act of speaking I exercise transeunt causal power of a complex sort on another person in the world about me.

But as I speak these words I also exercise another kind of power, at once more fundamental and more easily overlooked than transeunt causal power. It is the ontic power/causality immanent in my self-identity. I exercise it over my brain—indeed, over my whole nervous system and my body—both in the sense that I levy the events in my brain to be components of my self-identity and in the sense that my intent to convey something to you makes those events (some of which have not yet occurred) have a structure different from what they would have if I were intent on saying something else. And it is this exercise of immanent ontic power that makes possible my exercise of transeunt causal power on you. It is the paradox of self-identity that all those things subject to the ontic power of the superordinate primary being make a correlative ontic contribution to that self-identity; we may thus distinguish supervening, or superordinate, ontic power from conditioning, or subordinate, ontic power.

In all this it is the power or force of a rational speech act subsuming something distinct from it that is at the focus of our direct (and in this case reflexive) rational awareness. There are complexities that cannot be addressed in detail here. For one thing, the remarks about the role of the nervous system obviously also involve indirect knowing on a theoretic basis. Yet, even if we knew nothing of the nervous system, rational awareness of the exercise of the function of rational speech would carry with it the knowledge that something or other that is bodily and in important respects distinguishable from rational speech is subject to it. For another, the ontic causality in which an action exercises power on one of its subordinate events may include the power of reasons, and the power of reasons can then qualify the transeunt causal power exercised on someone else.

The power of reasons has an additional feature still more difficult to conceive of in terms of dominant philosophic theories of causality. It is present only in the kind of action I took as my example, action that issues in rational speech and is therefore characterized by both the function of rational awareness and the formative function of rationality. The variety of such actions is considerable. On the theoretic side, it ranges from the speaking or writing of sentences of modest import to the enunciation of important scientific truths and the construction of momentous scientific bodies of theory; on the practical, it ranges from the involvement of rational speech with the ordinary commonsense tasks of daily life to its involvement with moral decisions of the most important kind. If we suppose that this considerable range of acts not only is supported by neuronal activity but also exercises power over it, then in the course of the action what is understood, what is articulated, what, it may be, is looked to as a standard for action all contribute to the power exercised by the action over its supporting neuronal activity. Though the way I approach these matters is significantly different from the way of most recent writers on reasons and causes, it should be clear that what is articulated in rational actions is akin to what is usually dealt with under the rubric of reasons.

Here, in the full spate of philosophical assertion, I am checked, if not by my own reflection, then at least by your response. Surely it is not proper to distinguish the *I* from the body and all the events that go on in it while the *I* lives. Surely we have no right to distinguish something called an action from all the things that are happening in the body as the action goes on. The *I* would seem to be identical with the history of the assemblage of cells that include my neurons; my action, in the same way, would seem to be identical with everything that is happening in the body while I perform it.

Your imagined objection insists on the identity of the items I have been distinguishing from each other: the *I* from its body, the action from the multiplicity of neurons and other physiological units that carry out their functions in the course of the action. Your objection is to the implicit dualism: you suppose that what you call the identity of the discriminated things makes that dualism illegitimate. I concede to your imagined objection as long as you do not conceive of the identity as a physicalist (materialist) would conceive of it. There is indeed an identity of the two items discriminated in these dualisms, but it is different from what the physicalist intends and so is best introduced by contrast with physicalist identity. The clearest example of physicalist identity is the familiar theory of mind-body identity. Let us remind ourselves of its main lines before returning to the qualified identity I have in mind.

Those who argue for the theory have always conceded that the terms 'mind' and 'body' are distinct in meaning. The theory thus harbors a quasi-dualism: there is obviously no identity of the two terms and no identity of their meanings. When identity is considered to be a relation, it is usual to consider it a symmetrical one; in the present case, whether we are talking about terms or their meanings, there is no such symmetrical relation. Retaining the single quotation marks to indicate that we are still talking not about entities but about terms and their meanings, we may then say, " 'Mind' is not identical with 'body'. " What, then, is the point of the identity theory? It is usually said that the two nonidentical terms/meanings refer to but one referent. But that in itself does not give us an *identity* theory, as distinct from, say, a two-aspect theory. What is essential to the identity theory is that there is a radical difference in the way the two terms refer to the entity in question. One of the two terms, 'body', is understood to refer appropriately, adequately, and rigorously (the referent really is body). The other term, 'mind', is understood to refer in a way that is acceptable only to the easy-going spirit of common sense—only, in effect, at the level of appearance (the referent is not really mind, but rather body). We may then say, speaking now only of the supposed entity, "Body is identical with body."

No doubt those who propound the theory take it for granted that identity, insofar as it can be regarded as a relation, as it often is in mathematics, is a symmetrical identity. But in the present case, all this means is that the one referent, body, is symmetrically identical with itself, and that the term 'body' (or the meaning of that term) is symmetrically identical with itself; and that is not very illuminating. But the symmetrical nature of the identity relation is not what the theory is about. The theory is in fact a reductive (or eliminative) one: it is designed to call our attention to the supposed unproblematic and ontically homogeneous self-identity of the entity in question and to the supposed error of those who persist in seeing some form of dualism in that self-identity. I have argued against this form of identity theory several times in the past.[14]

My own way of discussing this question—the way I imagined you to be objecting to—was designed to focus on that primary being, the rational agent, and to allow our rational

14. See especially *Acts of Our Being*, chapter 6, pp. 168–90.

awareness to take hold of it precisely as it is. If we do that, I think we shall find that the self-identity of the rational agent is not in the least unproblematic and ontically homogeneous but rather paradoxical and ontically complex. In trying to draw attention to that ontic complexity in the past, I have regularly used the expression 'asymmetrical identity' about it, but that has troubled some otherwise sympathetic readers who take it for granted that identity is by definition a symmetrical relation. It would have been better if I had used the expression 'asymmetrical self-identity' instead. That expression, like the earlier one, is meant to call attention to the unity of a nature from which we can legitimately abstract various quasi-dualisms, each emphasizing some aspect of its complexity that happens to interest us.

None of these quasi-dualisms—the *I* and its body, the self and its body, the unity of an action and the multiplicity of sub-acts or other events we can find in it, and mind and body—is absolute, for none of the pairs consists of items that are absolutely independent of each other. Their interdependence is in fact so intimate as to suggest that some qualified sense of identity is at least as close to the truth of the matter as is the notion of a dualism. If, for instance, we distinguish the unity of a rational agent from the pattern of the firings of all its neurons in the course of one of that agent's acts, the relation between the things discriminated is nothing if not intimate. Let the macroscopic act differ in some way, and the pattern of firings differs too; alter the pattern of firings, perhaps by introducing a drug, and the macroscopic act alters—may indeed vanish. What the agent performs is performed in and with those firings, so it is easy to see why you might insist that 'agent' really means nothing more than an ordered complexity of many firings.

On the other hand, the identity cannot be absolute and unqualified, and it is certainly not ontically homogeneous in the sense intended by physicalist identity theory: there is *within* the identity something persistently quasi-dualistic. If the identity is not of the kind in which one of the two items is only spuriously distinct, as in the physicalist theory, neither is it an identity of coequals. If one sets about altering the pattern of neuron firing as such, one does so by drugs or by surgical or electrical intervention; if one sets about altering the act as such, one must do so by communicating with, or otherwise acting on, the agent as such. Moreover, although the self-integration of the agent (as one might call it in a moral context) may vanish under disease, drugs, or torture, self-integration of that kind is not automatically procured by the absence of those things. All this seems enough to make us want to keep at least the distinction marked by the two expressions 'agent' and 'pattern of neuron firings'. Although a given neuron may be inhibited from firing by the contribution of the neurons in synapse with it, the total pattern of neural activity to which those inhibitions belong is what it is because of the total form of what the agent intends to carry out and does carry out: thus I foresaw, when I began to say what I have just said, the general shape of it, though not perhaps these very words with which I now bring it to an end.

Perhaps the clearest contrast to be found in the asymmetrical self-identity that concerns us is that between the unity (or integration) characteristic of the rational agent and the vast multiplicity of other primary beings and other events held together in that unity. And there are many multiplicities besides that of neuron firings. But even the general contrast between a unity and a multiplicity is itself a kind of quasi-dualism; indeed, it is the very paradigm of ontic heterogeneity. If the asymmetrical self-identity of the primary being in question is acknowledged in somewhat this sense, we might then continue to speak of the members of each pair of features (or factors) we discriminate within it as asymmetrically identical with each other.

All these findings are in the long run findings made by rational awareness. Asymmetrical identity of the member of a certain ontic level with its infrastructure is precisely what we are rationally aware of when we are aware of a primary being. And while just here we have been concerned with the theme of asymmetrical self-identity, that theme is not radically different from the theme of causality; indeed, if the asymmetry were not also of causal significance, it would not be of much importance. When we attend to the asymmetry, we are also attending to the distinction between supervening and conditioning ontic power/causality and the distinction between ontic power/causality in general and transeunt causal power. The asymmetry appears in the ontic causality operative between any two ontic levels regarded as simultaneous. And because the transeunt causal power by virtue of which any entity is able to exert causal influence within its own ontic level—as persons may influence other persons or neurons stimulate other neurons—originates in its own superordinate ontic power, the asymmetry is a feature of transeunt causal power as well. Our attention moves back and forth between the ontic constitution of an entity and the causal significance of that constitution.

The adjustments in our habitual beliefs this doctrine of causality requires are not in conflict with the assumptions of science about transactions in which physical energy is lost or acquired. If we compare the transeunt causal power I exercise on your ears by way of sound waves with the ontic power I exercise on some neuron that fires or is inhibited from firing during one of my actions, we see a marked difference. The first is an ordinary physical transaction; it takes up time, and I, distinct from you, expend energy (in the ordinary physical sense) to agitate your tympanums. The second is not a physical transaction in that sense. If when I speak to you I exercise power over some neuron or neuronal complex, the power is not of the kind exercised by way of a physical transaction that takes up time and in which physical energy is expended in the production of a later effect. By the same token, the firing of the neuron or the coursing of impulses through a neuronal complex does not support the agent's thought and speech by expending physical energy on them, although it does expend physical energy on other neurons and neuronal complexes and hence does exercise transeunt causal power on them.

The whole-part relation of act (or function) to the firing of some single neuron or neuronal complex makes vivid the temporal distinction between this asymmetrical mode of interdependence and dependence in the mode of transeunt causal power. The activity of some single neuron is asymmetrically interdependent, in the mode of ontic power, with a macroscopic feature of a speech act or with an instance of rational awareness that embraces it temporally; it is thus subject to the superordinate ontic power of the act or of the exercise of the function. That neuron fires many times in the course of the time span of the act or the function. Its firing takes place within the total temporal pattern of the firing of billions of other neurons during the same time span. All those neurons are subject to the superordinate ontic power of the act or function. The act or function is asymmetrically identical with the firing of those neurons; they in turn make a conditioning ontic contribution to the superordinate, and both the supervening and the conditioning ontic power are simultaneous. But in the course of the development of the act or function we are considering, the particular neuron we have been singling out for an example also makes a contribution in the mode of transeunt causal power—indeed, many such contributions—to later neural occurrences within that same development but in some other part of the brain.

We cannot give a satisfactory account of ontic power/causality by analyzing it into the various cause-followed-temporally-by effect transactions (instances of causal power) that can be discerned within the complex or multiplicity asymmetrically identical with it. Nor can we give an adequate account of an instance of supervening ontic power by listing the various conditioning powers that make an ontic contribution to it. For that reason I borrowed, in *Meditation on a Prisoner*, the Platonic phrase 'real (or true) cause' to refer to the agent in its exercise of supervening ontic power, making it clear, however, that I was not borrowing all the other well-known tenets of Platonism along with it.[15] In responsible moral actions, and indeed in what I called in *Meditation on a Prisoner* originative acts in general, the agent itself is the real (or true) cause. By this I do not mean just that the agent is the real cause of what happens in the world as a result of its actions, although in the mode of causal power it is just that. I mean rather that it is the real cause, the unifying power, of the total act and thus of much in the complexity of non-acts into which analysis might seek to resolve the act. The expression 'real cause', then, is another way of calling attention to the supervening ontic power of a primary being.

It is obvious that I have tried to absorb and transform in the notion of real cause, or ontic power, certain aspects of causality that have long since disappeared from the notion of causality taken for granted in modern philosophy of science. As for causal power, the most important things to be said are that it is derivative from ontic power and that its exercise involves a temporal sequence. Thus one neuron, considered as an infrastructure unit within a superordinate primary being, exciting another with which it is in synapse, exerts causal power on the latter but does so by virtue of its ontic power; and the firing of the latter occurs after the stimulus, or in any case is completed after the stimulus. In earlier work, I have sometimes used the expression '$C{\to}E$ causality' about causal power to bring out the importance in it of temporal sequence.[16]

For primary beings that do in fact have infrastructures, the exercise of ontic power always means that the being is asymmetrically identical with its infrastructure. So, though

15. Pols, *Meditation on a Prisoner*, as indexed under 'causality, "real" '.
16. The working scientist generally disregards Humean objections to the authenticity of our knowledge of causal power, and so I distinguish the working version of $C{\to}E$ causality from the official version, which is the version propounded by philosophers who find Hume's account of causality persuasive; see Pols, *Meditation on a Prisoner*, pp. 16–27. That whole book is in effect a discussion of the distinction between causal power, or $C{\to}E$ causality (both official and working), on the one hand and ontic power on the other; for details, see the analytic index under the general heading 'causality'.

our neuron is a subordinate, it is a superordinate with respect to its own infrastructure, and its exercise of causal power depends upon that superordinate status; so in turn with *its* subordinates. Instances of causal power within its infrastructure—as when, for instance, we give an account of the firing of the neuron in terms of the passage of sodium and potassium ions through ion gates—are derivative from the ontic power of such subordinates. The principle is, then, a hierarchical one: any primary being that plays a role in some infrastructure and in turn has an infrastructure of its own is both subordinate and superordinate. It follows that an account of a superordinate unit in terms of an infrastructure understood in the light of a causal doctrine that neglects the dependence of causal power on ontic power/causality is a deficient one.

Meditation on a Prisoner was not designed to supply a complete first philosophy. It was meant to draw attention to a feature of action that calls for a recasting of many received ideas about causality and to some other features of action that call for a recasting of some received ideas on the role of mind in human action and human affairs in general. But I do not labor under the misapprehension that the human self can be adequately understood in terms of action alone. I tried to make that clear, in that book, by way of the notion of being that is present in the expression 'ontic power'. I also tried to make it clear that ontic power is possessed by all primary beings, most of which can hardly be said to act, except in some metaphorical sense. The focus of that book is different from that of the present one, but most of what is said there about action can be drawn on to augment the present account of the asymmetrical self-identity of the rational agent.

Returning to the ontic level of the rational agent—the level that exercises the function that is our theme, rational awareness—we should notice that the emphasis on the simultaneity of the relation between a superordinate ontic power and its subordinates must be balanced by an insistence that the unity of an action, which is a prominent feature of at least primary beings of our kind, is itself a temporal unity, although one neglected in the mathematical models of time that have dominated in science until quite recently. In the books mentioned, I called this unity 'act-temporality' and spoke of it as one of the many instances of the 'tensive' nature of time. If this is indeed time's nature, then time is more hospitable to a telic interpretation than has been generally supposed in this century.

Agents are, ontologically speaking, more fundamental than their acts. They are so fundamental that there is good reason to call them primary beings, although no reason at all to think that they alone deserve that name. As for the expression 'primary being' itself, it is, as noted earlier, a deliberate echo of Aristotle's *protē ousia* and a deliberate bit of propaganda against our continuing to translate that most important Greek expression by the totally inappropriate English expression 'primary substance'. In this book I want to make more forcefully a point I made in passing in the books I mentioned, namely, that causality itself is merely an aspect, although a most important one, of primary beings as we know them directly—an aspect of their self-identity and integrity, but an aspect also of their interdependence and togetherness. To know a primary being directly is also to know causality directly: to recognize the one is to recognize the other. To put the matter differently, our rational recognition of primary beings is the real warrant for us to discriminate, within their functioning, an ontic power that makes them capable of exerting transeunt causal ($C \to E$) power. For our rational recognition of a primary being is in effect our recognition of the superordinate level not as a mere appearance but rather as a center of ontic authority. With appropriate adjustments, all this is true about our recognition of artifacts as well; the ontic authority in that case is simply that of the agent who made the artifact.

A primary being, when recognized by our experientially engaged reason, has a unity that manifests itself, or expresses itself, in two ways, both of which are presupposed by the notion of ontic power. It is, in the first place, the unity of a temporal and dynamic thing—the unity of something that has a beginning, a middle, and an end. In considering the unity, we need not be disappointed at the absence of something that persists unchanged, as a substratum, through the change. The shaped and directed time of the superordinate level, which is the level toward which our rationality is directed, is not at odds with that unity; instead, it *expresses* that unity by being itself an ordered and directed *temporal* unity. Alternatively, the unity of the level in question realizes itself, actualizes itself, in the ordered change. The act-temporality of human acts—something I made so much of in *Meditation on a Prisoner*—is merely a special case of this. Thus the unity of each act, and the unity of a superordinate act embracing many sub-acts, also expresses the unity of that primary being the human agent. In the second place, that unity expresses or realizes itself no less in a spatial order, for much of its temporal dynamism consists in the development and maintenance of a relatively stable spatial structure, and all of its mature functioning takes place by virtue of that stable structure.

If this account is sound, it is obvious that in recognizing a superordinate primary being we have been attending to telic and formal features that we have not imposed by our attending. If we call attention to a primary being in this way, and do so with some reason, we are in fact calling attention to its explanatory value. To speak of its ontic power over subordinates with which it is asymmetrically identical is simply to draw attention to one feature of this explanatory value.

We have been concerned in this section with our direct rational awareness of causality. This is the source of our authority for formulating propositions expressing such causal principles as the "thin" version of the principle of sufficient reason considered in the previous section—or, for that matter, any "thicker" version one might venture to formulate. It is also our authority for rejecting a variety of propositions in which one might sum up claims that we have no cognitive access to a real causal relation. We need not, for instance, accept the claim that if we are indeed rationally aware of the cause in some exercise of causal power, we must also be aware, in advance of its completion, of precisely what it will necessitate as an effect in the world about it. That claim is not true, even though, when attending to one entity exercising causal power on another, we are in fact rationally aware of its necessitating *some* change in the other entity. We may thus say that our direct rational awareness of causality allows us to set aside these claims made by Hume:

> From the first appearance of an object we can never conjecture what effect will result from it. But were the power or energy of any cause discoverable by the mind, we could foresee the effect, even without experience, and might, at first, pronounce with certainty concerning it by the mere dint of thought and reasoning.
>
> *First*, it must be allowed that when we know a power, we know that very circumstance in the cause by which it is enabled to produce the effect, for these are supposed to be synonymous.[17]

And because rational awareness of the exercise of ontic power and of the dependence of causal power on it is also rational awareness of the connectedness of events that involve primary beings, we can also dismiss Hume's claims about the lack of connection between events:

> All events seem entirely loose and disconnected. One event follows another, but we never can observe any tie between them. They seem *conjoined*, but never *connected*.[18]

---

17. David Hume, *An Enquiry Concerning Human Understanding*, sect. 7, pt. 1, in Hume, *Essays Moral, Political, and Literary*, 2 vols., ed. T. H. Green and T. H. Grose (London: Longmans, Green, 1889), vol. 2, pp. 52, 56.

18. Ibid., p. 61.

4 The Exemplification, in Rational Awareness Itself, of Its Own Findings in First Philosophy

The conclusions of a first philosophy that knows causality and primary beings directly bring us right back to rational awareness, for rational awareness is a function of a primary being and so exemplifies the causality we have been discussing. It is an additional complication that our exercise in first philosophy is in fact a reflexive deployment of rational awareness, but this complication in no way invalidates our conclusions. Two aspects of causality are accessible to rational awareness, and these aspects yield two senses of the term 'causality'. The first and more basic, ontic power/causality, concerns the structure of a primary being and, because our example is a rational agent, the structure of the agent's acts and other functions as well; the second and more derivative one, transeunt causal power, concerns the efficacy of the primary being in the world around it, and therefore the efficacy of the agent's acts and functions in the world around it.

Consider now the function of rational awareness as it is exercised by a primary being like ourselves. This is by no means a call to introspection, for I mean both our particular selves and others around us who are just like ourselves and are seen to be so. They are primary beings, and in attending to them we take them to exemplify that group in general; but we are also attending to them as rational animals like ourselves. All that was said earlier in our consideration of primary beings and their causality still applies. Each such primary being has a functional, operative, powerful self-identity by virtue of the being's asymmetrical identity with all the subordinate primary beings that make up its infrastructure, where 'asymmetrical' means that its functioning as a powerful, superordinate self-identity requires the subsumption of those subordinates. Therefore when we, exercising our own self-identity in the function of direct knowing, recognize some other primary being, we must do so by subsuming, in order to attend to the being in question, what is really subsumed in that being itself. Subsumption of infrastructure elements that contribute to the self-identity of the superordinate being we focus on is of the nature of rational attention and its completion in rational awareness. The primary mission of this superordinate, qua knower, is to become rationally aware of—that is, to know directly—some other superordinate that falls within its limited capacity for direct knowing. And this is to say no more than that it is the mission of our rational awareness to take the directly knowable *as* superordinate if it is indeed superordinate.

Although we have been considering superordinates that are primary beings and thus exercise ontic power, not all superordinates are of that sort. Sometimes the knowable thing is a symbol. In that case the superordinate level of some mark on paper or some sound is its being such and such a symbol. It has indeed been constituted as a symbol by virtue of what I have been calling the second, or formative, function of our rationality. But our appropriate response to it is still that of superordinate to superordinate; one misses the reality of the symbol qua symbol if one takes it to be just a mark on paper or a sound. But in recognizing a mark as having been constituted as a symbol I do not, qua recognizing it, perform an act of constituting; I merely become rationally aware of what it in fact is—a mark or sound constituted as a symbol. The case of artifacts in general is similar and requires a similar response on our part.

In this radically realistic alternative to a Kantian synthesis, we are not synthesizing at all but rather responding adequately to what is already "synthesized"—to what is already a unity before we respond to it. The act of rational attention does indeed subsume infrastructure elements to know that superordinate directly, but it does so because they are already subsumed in the superordinate real being we are responding to, and not because it constitutes, or synthesizes, the superordinate out of elements which, in nature, are not under any higher ontic control. To know directly is to exercise a radically realistic power, however difficult such a power may be to credit, let alone understand.

Our experientially engaged rationality, confronting the temporal and spatial unity of a primary being, an artifact, or a symbol, thus responds to it, attends to it, acknowledges it, recognizes it. Rationality does not merely survey and summate a bare temporal and spatial extensiveness that is merely experienced and then accomplish a unification of that supposed booming, buzzing confusion by applying some concept, name, term, or category to it. It does not, that is, so organize an experiential welter—a welter that does not have "on its own" even the status of falling naturally into more or less discrete temporospatial nodes or events—that it appears as, is known as, something it in fact is not. Rational awareness, as distinct from a purely sensory apprehension—if indeed there ever could be such a thing for us—unfolds itself in an appropriate linguistic response in the presence of the unity of which ordered extensiveness is the expression. It is important that it responds to an operant unity and does not impose a unity on something that lacks it. The act of rational attention thus responds to, assimilates, acknowledges the asymmetrical "relation" between a superordinate and its infrastructure. And that is why, when it is engaged in the concrete problems of first philosophy after a reflexive turn, it can take the transeunt causal power exerted by a primary being on its environment to be derivative from, or an expression of, the ontic power/causality of the superordinate.

If rational awareness can attain primary beings and their causality not by propounding a theory about them but only by actualizing itself in their very ontic and causal structure, we are left again with a self-justifying function. We relied, in Chapter 5, on the rational-experiential satisfaction inherent in rational awareness to assure us that we are indeed capable of knowing directly things that are independently real. And now we rely on the texture of rational awareness itself not only to discriminate something important about the ontological and causal structure of those real things but to assure us that in fact it has done so. The known thing's independence of the formative power of rationality, together with the intimacy and familiarity of its assimilation in the act of rational attention, announces itself in the tissue of the act itself. We have no better warrant to offer than this, that rational awareness, so engaged, subsumes what is in fact subordinate under what is in fact superordinate—and that it does so because it recognizes this to be so. But it is only at the level of reflexive intensification that justification takes place, for at that level the satisfaction in the thing known, and in that only, becomes qualified by a satisfaction also in the generality of rational awareness-in-act and in its one inalienable birthright. That is its attainment of Being in general as independent of the act of direct knowing even when it is in doubt about whatever particular instance of Being purports to be present. Thus there is a reflexive knot at the heart of this example of a substantive problem of first philosophy, just as there was one at the heart of our very first step in first philosophy. But the obverse of this rather negative way of expressing that point is, first, that we have found a living rational authority working in first philosophy, and, second, that it suffices.

We have not, however, quite exhausted the intricacy of the knot, the fruitful reflexive closure of the function of rational awareness. Our capacity for achieving rational awareness of something that does not depend on the function of rational awareness for its ontic integrity, together with our capacity for reflexively intensifying rational awareness, is an aspect of our own ontic integrity as rational persons. We were talking just now of the way rational awareness deals with its objects; we must now notice how, precisely as a function, act, state, or condition of the knower, rational awareness itself exemplifies the distinction between superordinate and subordinate—exemplifies it in the very act of responding to that distinction to attain its object. Acts of direct knowing can take place only insofar as knowers subsume their own infrastructures in the act of knowing itself. The act of becoming rationally aware *is* such a subsumption: it exercises ontic power/

causality over the infrastructure it is asymmetrically identical with. It thus lives in and through our neural networks, but only insofar as billions of neurons are subsumed in—indeed, only insofar as they vanish in—the act of rational awareness itself. Our very capacity for direct knowing, including our capacity for the very reflexive intensification I have been trying to bring about, is part of a powerful, reigning unity or self-identity and so takes place only in virtue of such a subsumption.

5 The Reflexiveness of Rational Awareness and the Movement from Primary Beings to Being

Direct knowing, whether it is of other rational agents, the earth, animals, plants, and artifacts or of terms, propositions, and bodies of theory, is always naturally reflexive. Even before reflection is discussed or systematically undertaken, it already colors our rational awareness: our rational awareness of something already involves our distinguishing that something from ourselves attending to it. No doubt it is the nature of the something that preoccupies us, but on the other hand its nature becomes evident through both its distinction from other things and its distinction from our attending. Reflection is incipient in all rational attention, and that is why we can call attention to it and intensify it.

The point is even clearer when what we are attending to is a proposition, for we are surely aware of ourselves as rationally sustaining the proposition—sustaining it so that it may be considered—even while we are in fact considering it. Whenever we attend to anything rationally, *our* attending to it is part of our attending to *it*. The obverse of this is that *its* exaction from *us* of a faithfulness to *it* is part of its cognitive presence to us. Direct knowing is always colored by its own reflexive feature, and the intensification of reflection makes that more evident. It is not, I think, a different reflection that accomplishes this, but truly an intensification of what is already operant in our daily cognitive occasions.

The intensification leads us to a more adequate direct knowing of direct knowing itself, or, if you prefer, to a more adequate rational awareness of rational awareness-in-act. Although the intensification leaves our commonsense knowledge in place, it carries us well beyond the merely inveterate character of common sense, for it brings with it the realization that to know directly beings that are independently real is also to know directly their rootedness in a common Being. Our attention is directed toward them as things we ought to be cognitively faithful to. But the more reflexiveness intensifies, the more we are aware that what lays the obligation on us is not a collection of bare particulars but rather particulars that are participant in the universality and unity of Being. Any directly known particular is thus attained not just as something that may be brought under the universality of concepts and categories but as something that participates in an ontic universality that makes possible the derivative universality of concepts and categories. It does not matter whether we call it the unity and universality of the nature of things or the unity and universality of Being, or, as I have called it so often, the U-factor. It is in any event a common source of whatever power any particular being might exercise, no less than a common source of any generality a term or concept might possess. One of our more important encounters with it is its presence in every instance of rational awareness itself. Whenever we intensify the naturally reflexive feature of that function, we come to see that, although it is one's own particular reason that finds satisfaction in some particular rational-experiential engagement, this satisfaction is not possible unless we are also satisfying a more universal power in which our particular reasons participate. We are vicars for an authority that transcends all the instances of its deployment in our particular selves. Whether *what* we are rationally aware of or the *function* by virtue of which we are rationally aware is at the focus of our concern, the U-factor asserts that it is there, however inadequately seen, and it demands a more adequate response on our part.

To some readers, all this will seem hopelessly speculative. Yet I have been trying to make a case for a prospective first philosophy nonspeculative in its essence and tolerant of speculation only to the extent that speculation does not violate that essence. It should be recalled that the findings of a reflexive justification remain rooted, for all the emphasis on the U-factor, in the particular primary beings with which we began. It is rather our rational awareness of what a particular being *is*, and also of what a single instance of rational awareness *is*, that changes. Being in general is no longer a speculative presence: it reveals itself instead as a decisive feature of the real presence of particulars, a decisive feature of the rational awareness with which we deal with particulars, and a decisive feature also of our reflexive intensification of rational awareness.

Now scientists, or at least the best of them, take this rootedness of the particular primary being in the One—in the nature of things—quite for granted, even though they may be innocent of the kind of reflexive justification I am now engaged in. Einstein even speaks, somewhere or other, of the Old One. And they try to make their versions of the laws of nature express that stability and unity. But an appeal to law is not the only explanatory appeal. When a primary being is present to our rational awareness, it is present as a particular instance of the power and order of which it is still appropriate to say, with Parmenides, that *it is* and that *it is One*. It is the multiplicity of such particulars, each rooted in the nature of things but rooted at the same time in generation after generation of like particulars, that allows us to talk with good reason of levels of Being, of essences, of natural kinds and also of laws appropriate to such levels, essences, natural kinds, and so on. This view makes laws in general, and therefore also moral laws, ontologically less fundamental than the beings and Being on which their status as laws depends.

The ontic power I spoke of in the discussion of primary beings, the ontic power of anything we attend to as superordinate, is therefore an instance of the power of the unity of things: it is the unity of things operative in the finite particular. When primary beings are present in direct knowledge, they are rationally present as participant in that unity of things. Similarly, rational awareness, present to itself in reflection—though always on some experiential base—is present as rooted in that same unity. If this is so, the consequence is momentous for our earlier discussion of the relation of ontic power that exists between superordinate and subordinate primary beings. We need no longer be disappointed that one cannot feel or measure that power. Nor do we need to trouble about such criticisms as that ontic power, if real, must behave like a signal transmitted anomalously—transmitted, in fact, instantaneously. We need not trouble because the very notion of a signal does not apply to ontic power. For if we now consider the unity of an act, the unity of a sub-act, the unity of a superordinate act, or again, if we consider the unity of a superordinate being and the billions of unities of its subordinate beings, we suddenly realize that we are not dealing with a mere collection of discrete unities.

It has been the thesis of this book that ontic power, or causality, undergirds direct knowing itself. That is to say that direct rational awareness is an expression of ontic power, as well as that by virtue of which we recognize real beings and Being—and hence recognize ontic power itself. How can we account for, how explain, a function so central to what we are? To hope to explain it by trying to undo the ontic bond between the unity of a superordinate being and the multiple unities of its infrastructure is to hope to untie a world knot that occurs in vast numbers of cases and that, wherever it occurs, is itself explanatory. To "account for" direct knowing we must acknowledge that knot and leave it as it is. Its cognitive counterpart is an inner resonance between the unity of the being that achieves rational awareness and the unity of the being of which it achieves rational awareness. In both cases the unity is asymmetrically identical with a complex set of transeunt $C{\rightarrow}E$ transactions that take place within and between the infrastructures of knower and known. To try to tell the whole story of that cognitive relation exclusively in terms of those $C \rightarrow E$ sequences brought under covering laws (to suppose that one has untied the knot) is to miss the whole point of our most fundamental rational achievement. If radical realism is true, we must accept the scandal at the heart of it: the only way it can possibly work is by virtue of an inner affinity, an ontic resonance, between the superordinate that is the knower and the superordinate that is the known. Two great overstatements of traditional epistemology—that of the innateness of knowledge, that of the identity of knower and known—express this resonance, though by no means adequately. I have tried in this book to elicit a more adequate expression from a reflexive intensification of our function of rational awareness, the very function whose nature is my central theme.

Index

Agent, rational, 172, 174, 203–4
Antipodean realism. *See* Scientific realism
Antirealism: defined 6–7; and relativism, 9–16, 20–21, 119–21, 171–72
Appearances: and antirealism, 6–8; 109, 120. *See also* Phenomena
Aristotle: astronomical theory of, 148–50, 168; on first philosophy, 173, 176, 189–90; on *ousia*, 188–90
Art: and reality, 11–12; and commonsense experience, 118
Assertibility, warranted, 60, 62, 65, 85, 98–99
Asymmetrical identity (of a primary being and its infrastructure), 198–205
Asymmetrical self-identity (of a primary being), 198–200, 207
Ayer, Alfred J., 89

Being, 210–15; and term 'reality', 13, 28, 177; and U-factor, 145, 153, 165
Body of language-cum-theory. *See* Body of theory

Body of theory, 4, 7–8, 30, 91–118, 133, 136–38, 166–69; and epistemic triad, 83–87; as experienced in rational awareness, 159; and the formative function of rationality, 146–51, 166; as isomorphic with target structures, 147–51, 167–69; metaphysical, 172–73, 176; as saving phenomena, 148–50, 167–69; as rational pole of science, 31–33, 73–77, 128, 156–61
Boyd, Richard, 100
Brentano, Franz, 37

Cartesians, scandal of the, 133–34
Cassirer, Ernst, 110
Causality: of primary beings, 188–206; and principle of sufficient reason, 181–88; and reference, 93–94, 97–98
Causality, immanent. *See* Causality, ontic
Causality, ontic (ontic power), 191–206
Causality, transeunt. *See* Causal power
Causal power, 192–206

217

Index

Causal principles, 181–88
Churchland, Paul, 100
Coherentism, 89n2, 97
Common sense, 80, 157; and art, 118; language of, 93–97, 131; things of, 2, 6, 67n3, 79; world of, 149, 157, 160–62
Commonsense experience, 3, 44, 79–82, 87–89, 92–98, 104–19, 149, 157, 163
Commonsense knowledge, 173
Community of discourse, 14, 120–21
Concepts, 74, 78, 96, 137, 158
Conditioning ontic power, 192, 200. *See also* Causality, ontic
Consciousness, 3, 33, 35–36; and intentionality, 36–37, 99
Contradiction, principle of, 183–84
Convergence, 100–102

De Anima (Aristotle), 189
Deconstruction, 58
Descartes, René, 62, 64n2, 188–90; and rationalism, 88; on representative ideas, 23, 79, 134, 180; on representative (objective) reality, 134n6; on subjective (actual, formal) reality, 133–34n5; subjectivism of, 36, 65
Devitt, Michael, 59, 100
Dewey, John, 60
Discourse on Metaphysics (Leibniz), 186
DNA, models of, 147–51
Dualism, 196–98
Dummett, Michael, 59, 85, 91n3

Einstein, Albert, 76, 213
Empirical pole (of science), 31–33, 128, 155–61, 165–67
Entia rationis, 43–44, 78, 117, 146, 154, 166, 170–71
Epiphany, 33, 39–43, 80, 154, 156, 176, 179, 188, 191
Epistemology, 34–35, 48, 79–80, 106–7, 119–20, 130, 133n4, 143, 154, 159, 166, 215

Essences, Aristotelian, 133n3
Experience: commonsense (prima facie), 3, 44, 79–82, 87–89, 92–98, 104–19, 149, 157, 163; flexible Kantian loading of, 108–18, 120; and linguistic consensus, 62–63, 66–77, 83–121; pure, 32; strict Kantian loading of, 106–8, 115–18. *See also* Rational awareness
Experiential pole. *See* Rational awareness

Feyerabend, Paul, 104, 109
Field, Hartry, 59, 99–100
First philosophy: and foundation metaphor, 50–52; and realism-antirealism debate, 168; and reflexivity of rational awareness, 32–43, 144, 153–56, 172, 175–215; and the term 'metaphysics', 24–26, 173
Formal systems, 55, 62
Formative function of rationality, 1, 9–11, 42–43, 49–54, 72–77, 122–26, 141–51, 165–66, 169–70, 176–77, 179, 181
Foucault, Michel, 53
Foundationalism, 50
Freud, Sigmund, 53

Given: Myth of the, 68n6, 89n2, 93; reality as, 90, 109; two senses of, 66–67
Goodman, Nelson, 59, 71, 105, 110–11
Grand unified theory, 4, 7–8, 162

Held, Barbara S., 120n28
Hooker, C. A., 100
Hume, David: on causality, 206; on experience, 88; on impressions, 133n4; and logical positivism, 77, 105; and metaphysics, 22n1, 23
Husserl, Edmund, 37, 63–64, 99

Idealism, linguistic, 113–14
Identity, 198–207

Identity theory (of mind and body), 196–97
Immanent causality. *See* Causality, ontic
Indirect knowing: and theory, 9, 74–77; and the formative function of rationality, 146–51
Innateness, 165, 215
Intension, 70
Intentio, 36
Intentionality, 63–64, 94, 99, 140; distinguished from rational awareness, 36–37
Intentional properties, 99
Internal realism, 91–102
Interpretation, conceptual, 115–18
Irrealism. *See* Antirealism

Kant, Immanuel, 62, 77, 92–93, 164, 181, 189; and art theory, 118n27; on experience, 68n6, 105–18; on intellectual intuition, 112; and metaphysics, 22n1, 23, 114n24; on morals, 107–8n15, 114n24; on objectivity, 112–13; and phenomenology, 36–37, 64n2; Putnam on, 112–15; and scientific truth, 10; on synthesis, 208; on thing in itself and noumena, 112–14; on understanding and sensibility, 106–7, 112–13. *See also* Experience
Kantian loading of experience. *See* Experience
Kantian synthesis, 208
Kuhn, Thomas, 59, 71, 117n26

Language: commonsense, 6; conflation with theory, 5–9, 55–57, 84; formal, 136n7; formative power of, 6, 9, 11; natural, 56, 84, 136n7; observation, 109; ontology of, 6–8, 57, 94–97, 103, 131, 150; and reality, 47–50, 52–54, 83–121. *See also* Body of theory
Leibniz, G. W., 56, 182–90
Lewis, C. I., 110

Linguistic community, 14, 120–21
Logical constructions, 67n3, 88–89, 95

Manifest image, 92–93
Many Faces of Realism (Putnam), 111–15
Marx, Karl, 53
Materialism, 92
Metaphysical theories, 23–25, 173–74
Metaphysics, 99–100; meaning of the term, 24–25; speculative, 25, 173–74; systematic, 173. *See also* First philosophy
Metaphysics (Aristotle), 26
Mind-body problem, 196–203
Mirror metaphor, 180
Models, 147–51, 158, 169
Monadology (Leibniz), 183, 186

Neo-Kantianism, 110
Neopositivism, 100–101
Neurath, Otto, 67n3, 89n2,
Nichomachean Ethics (Aristotle), 189
Nola, Robert, 59, 100
Nonrealism. *See* Antirealism

Observation statement, 85n1, 88, 89n2
Ontic level: of common sense, 31–32, 169; of the person, 8, 20, 145, 158–59; of rational agent, 172, 203–4; of rational awareness, 33
Ontic power, 191–206
Ontology: of a doctrine, 43; of a language, 6–8, 57, 94–97, 103, 131, 150; traditional sense of, 8, 130, 172, 177
Ousia, 188–90

Particularity, 37–38, 142, 156, 178–79, 190–91
Person, 8, 20, 145, 158–59. *See also* Agent, rational
Phenomena, 104, 107–8. *See also* Appearances; Kant, Immanuel

Phenomenology, 36–37, 63–64, 99
Philebus (Plato), 184–85
Philosophical Investigations (Wittgenstein), 105
Physicalism, 93–94, 97–99, 196–99
Plato, 12–13, 28, 132–33, 182–85, 201–2
Platonists, scandal of, 132–33
Positivism, logical, 67n3, 88–89, 95
Primary beings, 188–206
Projection, perceptual, 110–11
Propositional reality, 117, 120. See also Reality: Pickwickian sense of
Propositions, 37, 47–49, 51, 60–82, 85, 99, 128–30, 136–39, 141, 143, 179; and the formative function of rationality, 144–47, 165; as rational-experiential, 188. See also Body of theory
Protagoras, 13
Ptolemaic astronomy, 148–50
Putnam, Hilary, 59, 94, 98–102, 105, 111–15

Quine, W. V. O., 59, 71, 89n2

Rational awareness, 3, 31, 45, 172; and causality, 190–211; experiential pole of, 38–39, 124–25, 128, 155–56, 159–60; and first philosophy, 144, 153, 175–215; and formative function of rationality, 144–46; primary mode of, 126–36, 140–41, 149; rational pole of, 38–39, 128, 159–60; reflexivity of, 32–43, 46, 65, 125, 128, 139–44, 172, 190–91, 195, 207, 210, 211–15; secondary mode of, 136–39, 141. See also U-factor
Rational consciousness, 3, 33, 35–36
Rational-experiential engagement, 3, 33, 35, 130, 138, 140, 176
Rational-experiential satisfaction, 41, 139–44, 176, 179, 209
Rationality. See Formative function of rationality

Rational pole: of rational awareness, 38–39, 128, 154–56; of science, 31–33, 73–77, 128, 156–61
Real cause (Plato), 201–2
Reality: and art, 11–12; commonsense, 33; and epistemic triad, 83–87; as extratheoretic/extralinguistic, 8; nonlocal, 76; Pickwickian (antirealist) sense of, 12–13, 48, 71–72, 94, 103, 111, 169; Plato's doctrine of, 12–13, 28; rational awareness as feature of, 22–23, 178–79; and verb 'to be', 13, 28
Reasons and causes, 195–96
Reference, 62, 70, 85, 96–102; causal theory of, 93–94, 97–98; and mind-body identity theory, 197
Reflexivity: and first philosophy/metaphysics, 20–24, 32–43, 175–215; of rational awareness, 32–43, 46, 65, 125, 128, 139–44, 190–91, 195, 207, 210, 211–15. See also Epiphany
Relativism: and antirealism, 9–16, 20–21, 119–21, 171–72; and deconstruction, 58; linguistic, 10, 119–21; and moral choice, 11; in psychotherapy, 120n28; and science, 10, 119–21
Representation, 75, 78, 93, 180. See also Descartes, René
Richards, I. A., 118n27
Rorty, Richard, 59, 68, 105

Saving the phenomena, 31, 104, 149–50, 167–69
Schlick, Moritz, 67n3, 88, 89n2
Scientific realism: and linguistic consensus, 7, 91–102; and relativism, 14–16
Scientists, scandal of the, 134–36
Self-deception, 169, 171
Sellars, Wilfrid, 59, 68n4, 89n2, 92–93, 105
Sophists, 13
Spectator theory, 42

Speculative-empirical cycle. See Theoretic-empirical cycle
Stimuli, 6, 42, 46, 70–72, 130, 163–64
Strawson, Peter, 22n1, 59, 105
Structuralism, 57–58
Subject, knowing: and Kantian objectivity, 112; and language-cum-theory, 8; and linguistic consensus, 64–65, 94–95
Subjectivism, 36–37
Subjectivity, 134nn5, 6
Subordinate ontic power, 192, 200. See also Causality, ontic
Substance, 181, 184, 188–90. See also Primary beings
Sufficient reason, principle of, 181–88
Superordinate ontic power, 192, 200. See also Causality, ontic

Tarski, Alfred, 85
Theaetetus (Plato), 13
Theoretic-empirical cycle, 32–33, 73, 82, 156–63
Theory: conflation with language, 5–9, 55–57; and epistemic triad, 83–87; epistemological, 34; grand unified, 4, 7–8, 162; philosophical, 23–25, 29, 74. See also Body of theory; Formative function of rationality
Thing in itself, 93, 112–14
Timaeus (Plato), 182–83
Transcendence, 40
Transcendental Ego, 40
Transeunt causal power. See Causal power

U-factor, 28–29, 40–41, 127–28, 145, 176; and causality, 190–93; and rational awareness, 123, 142, 153, 156, 165, 212–15
Unity. See U-factor
Universality, 39; of Being, 145, 153; and rational awareness, 123, 142. See also U-factor
Universals, 38, 137

Ways of Worldmaking (Goodman), 110–11
Whitehead, Alfred North, 173
Wigner, Eugene, 81
Wisdom, John, 22n1
Wittgenstein, Ludwig, 53, 105
Worldmaking, 71, 77, 110–11, 117–18

Library of Congress Cataloging-in-Publication Data

Pols, Edward.
　Radical realism : direct knowing in science and
philosophy / Edward Pols.
　　　p.　cm.
　Includes bibliographical references and index.
　ISBN 0-8014-2710-X (cloth)
　　1. Realism. 2. Science—Philosophy. 3. Common sense. 4. First philosophy.
I. Title.
B835.P65 1992
149'.2—dc20　　　　　　　　　　　　　　　　　　　　　　　91-55531